THE BIM MANAGER'S HANDBOOK

Dominik Holzer

THE BIM MANAGER'S HANDBOOK: GUIDANCE FOR PROFESSIONALS IN ARCHITECTURE, ENGINEERING, AND CONSTRUCTION

Dominik Holzer

WILEY

ISBN 978-1-118-98242-6 (hardback); ISBN 978-1-118-98240-2 (epdf); ISBN 978-1-118-98234-1 (epub); ISBN 978-1-118-98231-0 (Wiley Online Library)

Executive Commissioning Editor: Helen Castle

Senior Production Manager: Kerstin Nasdeo

Assistant Editor: Calver Lezama

Cover design and page design: Artmedia

Layouts: Aptara

Front cover image: Copyright © Morphosis Architects

To my architect wife,
who doesn't understand BIM.

INTRODUCTION: WHY BIM MANAGERS COUNT!

BIM is changing, and rapidly so. While it remained predominantly the domain of technology specialists in architecture and engineering firms in the early twenty-first century, it is now steadily gaining relevance for a broad range of stakeholders in the design, construction, manufacture, and operation of built assets. Hand in hand with the dissemination of BIM comes the dissemination of knowledge associated to its application and the diversification of tasks associated to its management. BIM Managers are becoming far more relevant than simply acting as implementers of technology. They are in fact change agents and if they do their job well, it ties in closely with the core business pursued by their organizations. Beyond that, BIM Managers are becoming key innovators who help to transform the construction industry and associated professions globally.

Figure I–1 University of Sheffield Heartspace, Sheffield, United Kingdom.
Copyright © Bond Bryan Architects LTD

This *Handbook* was conceived to offer concise guidance and support to those trying to embrace the many facets of BIM Management. The chapters herein were originally published online as six eParts, each one related to all others, but at the same time sufficiently distinct to act as independent contributions to a whole. The sequential release as eParts has led to discrete, easily digestible sections on highly profiled topics, allowing for latest trends and developments about BIM to be included. In book form, the structure has the advantage that contents are very focused. The reader can go to individual chapters on a needs-to basis for information and advice.

The BIM Manager: Focus on the Person behind the Title

This publication adds to the existing body of work about BIM by taking a specific stance, namely the view of the BIM Manager. *The BIM Manager's Handbook* not only offers insights into contemporary research and trends associated to BIM, it is also highly reflective about the opportunities and challenges related to work undertaken by BIM Managers in contemporary practice. Over 50 leading architecture, engineering, and construction experts from the United States, Europe, Asia, and Australia have lent their voice in telling their stories and providing their feedback to this publication. Their view is that the job title of "BIM Manager" cannot easily be identified via a uniform set of tasks. Instead, BIM Manager roles vary greatly across sectors and companies. Clearly falling under the emerging field of Design Technology, BIM Manager tasks stretch across a great number of responsibilities associated to the planning, design, delivery, and operation of built assets.

Channeled into six cohesive chapters, *The BIM Manager's Handbook* offers a key reference for those currently engaged with BIM—as well as those who are considering applying BIM on future projects. The chapters put equal emphasis on practical application as well as strategic planning and overarching principles associated to implementing BIM. One other factor that sets *The BIM Manager's Handbook* apart from related publications is the fluent cross-over of technical, social, policy, as well as business-related aspects of BIM. The role of the BIM Manager is in constant flux. BIM Managers stem from all walks of life: technology gurus, 3D modeling specialists, construction experts, drafting guns, coordination experts … the list goes on. In current practice, most of these self-proclaimed BIM Managers have somehow grown into the role with only a very small percentage having undergone specific BIM Management training.

Given the ever-expanding context of BIM, one might struggle to find a clear definition of what BIM Managers actually do. Yet, despite the multiple directions in which to respond to this question, the answer is simple: BIM Managers are here to *manage*. They manage process, they manage change, they manage technology, they manage people, they manage policies and in doing so, they manage an important part of their organization's business.

Paradoxically, as representatives of a newly emerging profession (if one can speak of one) BIM Managers are rarely skilled in management. More often than not, they are tasked to perform a narrow set of practical tasks that respond to day-to-day affordances of practice. If in the past it was sufficient for BIM Managers to know their tools, workflows, and workarounds (combined with decent people skills), the property, construction, and design industries start to expect more: With the increasing understanding that BIM is not merely a technical

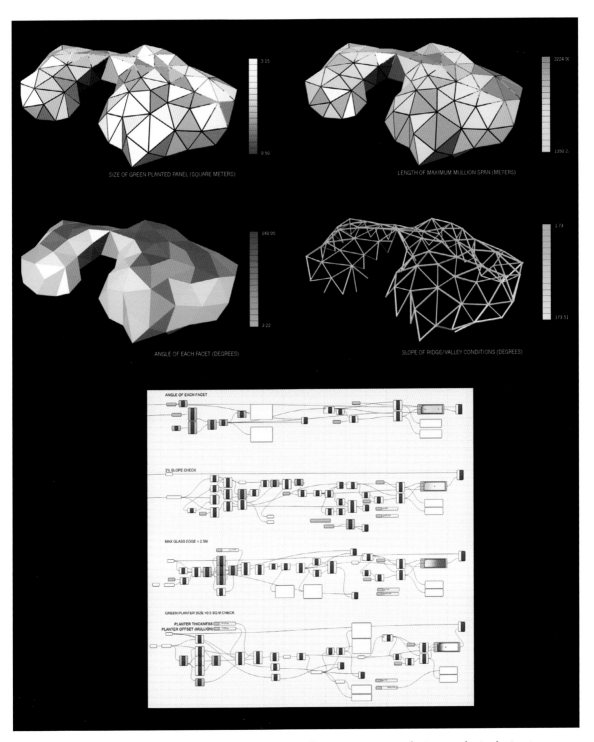

Figures I–2 through I–4 KAFD Conference Center parametric design analysis, design to structural node fabrication, and subpanel layout.

Copyright © Skidmore, Owings & Merrill, LLP

Figures I-2 through I-4 *(continued)*

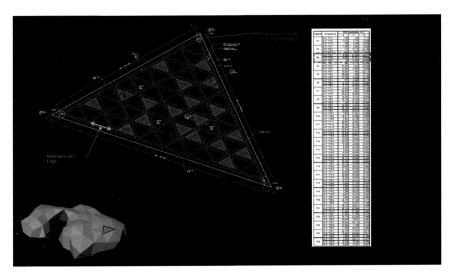

Figures I–2 through I–4 (continued)

side aspect of practice comes the expectation that BIM Managers need a broader set of skills including "management acumen." Such expectations are not only tied to a sound business sense, but they equally respond to an ever-growing set of policies, standards, and in some cases "mandates" that address how BIM is to be delivered in local jurisdictions. BIM is now more widely acknowledged as a contributing factor to reduce waste, the initial cost of construction, and the total cost of ownership of built assets. Next to that BIM can help to increase productivity across the construction supply chain and to reduce the impact of construction on the environment.

Figure I–5 Project coordination meeting based on BIM.
Copyright © Point Advisory

Given these realizations, it is surprising that the activities undertaken by BIM Managers are often badly understood within their organizations. It is not uncommon for BIM Managers to be tasked in defining their own role and to justify to upper management what it is they do.

Anyone trying to draw a precise boundary around the role description of a BIM Manager will soon realize the pointlessness of such an attempt. Roles depend on the tasks at hand and the distribution of responsibilities across multiple stakeholders. With any new job these tasks and responsibilities change and so does the role of the BIM Manager. In addition to the dynamic inherent to BIM Management, it is more than likely that in the future BIM will form an integral part of project design, delivery, and the operation of built assets. Its application will cease being looked at as a separate component and those we currently identify as BIM Managers will simply be "Designers," "Engineers," "Contractors," (or others) without requiring a BIM label. For now, BIM is still going through different rates of adoption throughout different industry sectors and geographic locations. Understanding its impact and the changes it effects on traditional means of project delivery is a crucial step for organizations to master. The BIM Manager(s) assist them on this path and they will do so for at least five to ten years to come.

Hands-On BIM

Instead of trying to offer an all-encompassing framework, *The BIM Manager's Handbook* explains how BIM can best be implemented by tapping into the on-the-floor experience of contemporary practice. By drawing from such expertise, hands-on feedback will guide the reader through a great number of real-life examples and anecdotes that will advance their own thinking. Many of these references get consolidated and summed up as practical "tips and tricks" that are easily digestible and translate to a great number of applications. Core to the information provided in all six chapters is the value proposition related to BIM and, inherent to this, the value proposition of the BIM Manager. The question thereby does not revolve any longer about use BIM or not, but about how to implement it successfully.

This *Handbook* clearly acknowledges the transient nature of BIM Management. It offers the reader an overview that aims at standing the test of time. The six chapters of this book each tackle a highly relevant portion of what those who manage BIM ought to know. First they set the scene on how to define Best Practice BIM in order to highlight the breadth of roles and responsibilities associated to its management. Drawing from this initial assessment, the consequent chapters then tackle distinct aspects of BIM Management in greater depth. Most importantly, this doesn't occur in the form of a mere technical explanation of day-to-day tasks. Instead *The BIM Manager's Handbook* addresses the wider significance of BIM Management responsibilities with far-reaching reflections on social issues, business directives, and knowledge acquisition. The reason behind this approach is simple: to answer what a BIM Manager needs to know and do in order to excel in his or her role.

When considering the BIM Manager role—the immediate needs and future requirements—it becomes apparent that there has been an overemphasis on the technology aspect in available literature. In response, this book only contains one chapter with a clear focus on technology. All others unravel the intricacies associated with

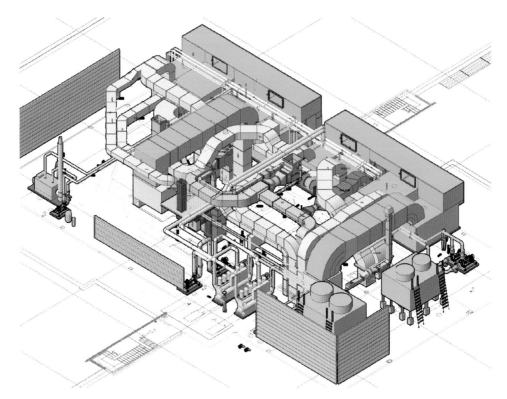

Figure I–6 Mechanical system plant room in BIM.
Copyright © A.G. Coombs Pty Ltd

BIM when instilling new ways of working, changing process, the importance of good communication, and the need for continuous skill acquisition.

BIM Managers need to learn to elevate their role beyond project support and position their activities among the leadership level of their firm and beyond. *The Handbook* describes how to achieve this change of emphasis and how BIM Managers can impact on the ongoing development of BIM through the construction sector, pushing for innovation and increased productivity. A great number of BIM experts and innovators who contributed to this publication are doing exactly this: sharing their research and facilitating dialogue with a high public profile. Examples of such excellence can be found in the work of Rob Jackson at Bond Bryan Architects in the United Kingdom who, together with his collaborators, keeps on investigating processes of IFC and COBie integration to the typical project delivery workflow. As a leading Quantity Surveyor, David Mitchell at Mitchell Brandtmann in Australia regularly publishes works about BIM "Return on Investment" on a macroeconomic scale. Another outstanding BIM proponent is James Barrett in the United States, who continuously reports on the approaches taken by Turner Construction to thrive for excellence in delivering projects using BIM and Lean Construction at conferences nationally and internationally. These are just some examples; more than 50 others have lent their voice to this publication.

In order to capture the knowledge of these global industry leaders, writing this book has taken the author on a journey of discovery and consolidation. An industry leader in his own right, who is actively engaged in the

Figure I–7 Ecclesall Road mixed-use development, Sheffield, United Kingdom.
Copyright © Bond Bryan Architects LTD

delivery of projects as well as the formulation of government policy surrounding BIM, it was pivotal to the author to reach out to a group of outstanding individuals globally. This exercise was undertaken in order to canvass both quantitative feedback (in the form of an initial survey), as well as qualitative comments from trusted experts in their field. The author has continuously expanded his network both geographically as well as thematically in order to capture both the breadth as well as the depth associated with BIM Management and its future development. The research for *The BIM Manager's Handbook* was undertaken online and in one-on-one encounters over a 15-month period. It resulted in numerous contributions from a lively, curious, and generous BIM community, united by common goals and concerns. One aspect that clearly emerged from the many discussions and the associated correspondence is everyone's enthusiasm and willingness to share their expertise and opinions. The insights offered in this publication are a testimony to the fact that the BIM community is bent on advancement over individual ownership.

Revelations and Surprises

One of the revelations from writing this book has been the startling low level of awareness of BIM Managers about their role within professional practice. With most of them entering the BIM domain via narrow pathways, they first need to expand their focus in order to understand the bigger picture. Even if they do, they then require convincing arguments to bring their firm's leadership on board and guide them in their process of making BIM

Figure I–8 Discussing latest software applications at a Revit Technology Conference (RTC).
Copyright © RTC Events Management Ptg LTD

work both internally, as well as across collaborative project teams. Providing such guidance does not come without a struggle: In a risk-adverse industry with low profit margins, the push for innovation and process change needs to be well orchestrated. BIM Managers are the key facilitators for change. They balance and harmonize the cultural with the technical, the business drivers with new opportunities of information transfer and sharing, the big policies with practical execution within teams.

The BIM Manager's Handbook demystifies a great number of misconceptions about BIM and it goes straight to the core of analyzing opportunities and challenges associated with BIM Management. It offers advice to fast-track every BIM Manager's and Design Technologist's development based on the knowledge of the best in business. It is set to become one of the key points of reference that will help us to globally take BIM further.

BEST PRACTICE BIM

How does one get Building Information Modeling right in practice? What are the key tasks and challenges faced by BIM Managers in achieving "Best Practice BIM" and how can they master them? By drawing from the experience of some of the world's top BIM Managers, this publication gets to the bottom of these questions. There is much we can learn from their experience, no matter if good or bad. The following exposé consolidates a broad range of feedback from these leading experts and it provides support to those who strive for excellence in their pursuit of implementing BIM.

If we want to understand how BIM Managers can excel in their role, we first need to understand the principles behind getting BIM right. This publication scrutinizes BIM's changing context and looks to see if there is a "BIM formula of success." The past decade has given us the opportunity to see a number of high-profile BIM projects through to completion. We learn from the mistakes we made on the way and we reflect on "Good," or even "Best Practice" BIM. What might be the tipping point for its successful implementation? What are the typical thresholds and benchmarks that apply? Answers to these questions will assist BIM Managers to maximize BIM benefits not only intraorganizationally, but also across the broader project team.

BIM Managers: Breaking Ground

BIM Managers are a wholly new breed of professional. They emerged internationally in less than a decade, most markedly in larger tier 1 architecture and engineering practices. By strengthening integration across disciplines and project phases, BIM Managers become the conduit for facilitating the information exchange between the design, delivery, construction, and operation of projects. They play a central role in deciding where BIM is heading. On a practical level, BIM Managers are the custodians responsible for innovation to occur within their organization and in collaboration across project teams. They empower project stakeholders to understand and engage with the high level of complexity associated with a BIM workflow. They help them to align their skills with the added benefits offered by data-centric and rule-based delivery of projects.

A Role in Transition

Describing what BIM Managers do is a difficult task. What was once associated with responsibilities for overseeing BIM model development is now more and more associated with information management, change facilitation, process planning, technology strategies, and more. Such is the veracity and speed of development surrounding BIM that the job description of any BIM Manager is in constant flux. Given the ever wider group of stakeholders BIM encompasses, there exists an increasing fragmentation of the BIM Manager's role into specialized responsibilities: On one end of the spectrum the role of Model Managers emerges, who assist in-house teams on individual projects, at times complemented by specialist BIM Librarians (or Content Creators). On the other end of the spectrum, Model Coordinators specialize in the oversight of the multidisciplinary integration of BIM. BIM Managers may now also report to Design Technology Leaders or Project Information Managers who directly report to upper management. In some instances, an organization calls for a Strategic BIM Manager (as opposed to providing more technical support on the floor). All of the above descriptions depend on the size and characteristic of an organization. In smaller companies, the BIM Manager may well be tasked to incorporate all those roles, while acting as Project Architect and BIM Modeler at the same time.

There is likely to be a time where BIM Managers become obsolete and their responsibilities will become part of project management in general. A good number of Change Management activities will have been implemented and construction industries globally will accommodate BIM as a matter of course in their project delivery methods.

For now, we still go through a major transition in adopting BIM. BIM Managers need to balance between the possible and the appropriate. Their strategic view will influence which opportunities can and should be aligned with the cultural and professional context of their organization. They also help to map out how such alignment can be achieved. In the end, BIM Managers may not be the ultimate decision makers in facilitating change. They are the ones who provide upper management with decision support in order to do so and they are the ones accountable for BIM implementation "on the floor."

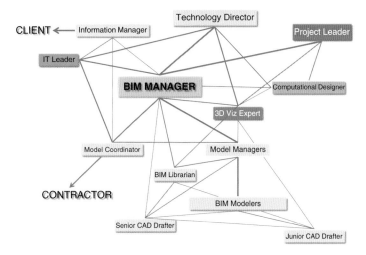

Figure 1–1 Mapping out a possible role distribution surrounding BIM in a larger size design firm.
© **Dominik Holzer/AEC Connect**

What makes a good BIM Manager, or even an outstanding one? In order to answer that question feedback is consolidated here from the world's top BIM Managers to make it accessible to everyone. These managers work for leading Architecture, Engineering, Quantity Surveyor (Cost Engineers or Cost Managers in the United States), and Construction firms. They report on pitfalls and the common mistakes associated with BIM to then highlight what makes BIM tick in practice.

The Rise and Rise of BIM

BIM use has been expanding continuously since 2003,[1] making BIM Management a moving target. Back then, BIM became the accepted industry acronym for a range of descriptions such as Virtual Design & Construction (VDC), Integrated Project Models, or Building Product Models. Until that point, different software developers had branded their tools with these varying acronyms, while essentially talking about the same object-oriented modeling approach that was first introduced to a wider audience by Chuck Eastman in the mid-1970s. Around 2002–2003, it was AEC Industry Analyst Jerry Laiserin[2] who played a pivotal role in promoting the single use of the acronym "BIM" which had been coined by G.A. van Nederveen and Tolman in 1992[3] and which later became the preferred definition of Autodesk's Phil Bernstein. It was the starting point for an industry-wide journey to holistically address planning, design, delivery, and operational processes within the building lifecycle. This journey raises a great number of culturally sensitive and professionally relevant issues: By nature a disruptive process, the adoption of BIM overturns decades of conventions related to the interplay between architects, engineers, contractors, and clients. BIM Managers are drawn right into the center of these changes in practice.

Despite the clarity about BIM's origin, there is no clear starting point to the commercial breakthrough of BIM; conceptually, BIM dates back to the early 1970s with the introduction of mainframe computers.[4] Some of the key BIM software platforms in use today have their origins in these early developments. The increase in processing

Figure 1–2 Detailed facade systems generated via BIM and visualized as a 3D rendering by COX Architects.

COX Architects

power, the drop in price for computer hardware, and the connectivity offered via the World Wide Web gradually led to an increased adoption of BIM in the early 2000s. During that period, a critical mass was reached. BIM software became affordable and it matured to the point where its user-friendliness offered a viable alternative to existing CAD platforms. From that point onward CAD Managers were those individuals most likely tasked with the oversight of the implementation of BIM. With documentation output in mind, CAD Managers were supported by senior drafting personnel who were responsible for generating the contractually relevant 2D plans/sections/elevations from virtual models. The process of BIM modeling remained limited to Architectural Designers and Structural Engineers. The limited scope of BIM existed much to the frustration of Services Engineers and Contractors who had to wait for the availability of BIM tools to serve their purposes until 2007–2008. From 2010 onward, developments surrounding BIM accelerated. Increased software interoperability and an ever-expanding BIM tool ecology resulted in BIM becoming more and more accessible to Quantity Surveyors, Contractors,

Facility Managers, and Client Representatives. The ever-expanding list of BIM stakeholders introduced a plethora of opportunities to manage information across disciplines and project stages. Significant consequences followed from this development.

With the broadening scope of BIM comes a diversification of what BIM Managers do: The more information can be exchanged by various stakeholders, the greater the possibilities and challenges for managing that information across those stakeholders. This expansion in scope has by no means occurred in a well-orchestrated fashion. On the contrary, it has evolved organically at different speeds and veracity throughout markets and industry contexts. In some cases there now exists a level of regulation about how information gets shared via mandates or incentives, in other cases the evolution of BIM depends on client demand or simply on the skill level of operators.

One commonality among these diverse propagations of BIM is the fact that until now, there has never been a clear educational pathway toward becoming a BIM Manager. When asking BIM Managers about their background at any conference, seminar, or local user-group session, they will likely represent a broad range of professional affiliations: (Recovering) Architects, Engineering Drafters, Quantity Surveyors, Project Managers, Service Contractors, Specialist Consultants—just to name a few. Some of these experts are self-taught and they have picked up their skills vocationally; others may have attended specialist courses or were introduced to BIM as part of their tertiary education. Others may have learned about BIM from colleagues in practice, and some simply may have picked up BIM as an expansion of the documentation processes they were used to from 2D/3D CAD.

From the early 2010s onward a number of professional bodies and academic institutions have started to offer tiered BIM Management courses with accreditations or certifications. Such courses denote that there exist fundamental, overarching themes that can be addressed in the context of BIM Management. The Singaporean BCA began their local BIM certifications in 2011–2012 as part of their BIM Academy.[5] Around the same time, the HKBIM in Hong Kong introduced entry requirements for their membership.[6] The Associated General Contractors of America (AGC) started their BIM education program[7] with a Certificate of Management—Building Information Modeling (CM-BIM) in 2011–2012. More recently, the UK–based Building Research Establishment Limited (BRE) announced a BIM training and certification pathway that focuses on the UK mandate that targets BIM Level 2[8] proficiency of stakeholders by 2016. What sets the BRE[9] approach apart from others is the split between Task Information Managers (TIM), Project Information Managers (PIM), and Project Delivery Managers (PDM). Less comprehensive, but with global outreach, is the BIM Manager accreditation introduced by the RICS in late 2013–early 2014; it predominantly addresses BIM Management for Chartered Surveyors, but accreditation is provided globally (albeit referring predominantly to a UK BIM context). The Canada BIM Council, CanBIM[10], joins the ranks of other industry bodies by establishing a Certification Program to provide: *A benchmark for individuals to be certified to nationally standardized and recognized levels of BIM Competency and Process Management.*

All of the above courses and accreditations were established by their respective industry bodies within the four years or less leading up to the first release of this publication. Many more are likely to follow. It is fair to assume that few, if any, of the BIM Managers who offer their feedback in this publication gained their knowledge from these courses. Yet this type of accreditation will become increasingly relevant for the second and third generation

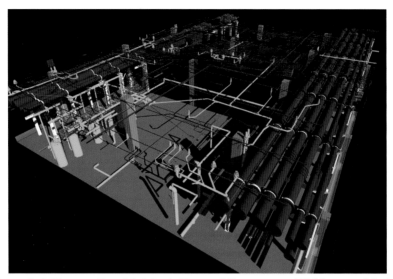

Figure 1–3 The new Royal Adelaide Hospital Construction BIM Services Model by the Hansen Yuncken Leighton Contractors Joint Venture.

© Hansen Yuncken Leighton Contractors Joint Venture

Figure 1–4 The new Royal Adelaide Hospital Field BIM used by the Hansen Yuncken Leighton Contractors Joint Venture.

©Hansen Yuncken Leighton Contractors Joint Venture

of BIM Managers to follow. How far the essence of BIM Management can be taught in class remains to be seen. BIM accreditation is without doubt an important stepping stone in order to address the epistemological aspect of BIM.

Defining what ought to be known in the context of BIM leads to a clear articulation of competencies and skills to be had by BIM Managers.

Defining Good, or Even "Best Practice," BIM

The term "Building Information Modeling" has remained of such a generic nature that interpretations about its meaning are vast and many. Some see "Modeling" as a verb, describing the activity of generating, assembling, and coordinating virtual building information.[11] Others refer to BIM as "a model" of building information, either in terms of geometric components, data, or a mix between the two. Considering the vast differences in defining BIM itself one needs to wonder if it is possible to define good BIM, or even "best practice" BIM.

The Big Picture

In some cases these documents lead to the generation of national policies or even mandates. An example of such guidelines is the UK Publicly Available Specification PAS 1192 with all its components and additions; another example is the State of Ohio BIM protocol.[12] These documents present the bigger picture of the aspirations related to BIM in local construction industries. They provide useful frameworks and a point of orientation to work toward for practices within a local industry context.

A semantic approach to any BIM definition is better left to the theorists. The work presented here is far more interested in the applicability of BIM as it unfolds in everyday practice. For that reason, this publication focuses on practical outreach and the application of tried and tested approaches to implementing BIM by drawing from the experience of leading BIM Managers around the world. It reports on cutting-edge research and practical use that helps to maximize the results of BIM-enabled workflows. Getting BIM right can never be a linear process as BIM is an ever-moving target. Well implemented BIM always relates to the combination of attitude/mindset and approach to the management of information across collaborators in general. Any attempt to defining Best Practice BIM needs to take into consideration BIM's transformative character that influences the array of stakeholders affected by its application.

We learn from examples and, when talking about the BIM, those examples often reveal a breadth of issues that cut through different, professional, cultural, and market-related contexts.

Reporting from the Trenches

When German Formula 1 driver Sebastian Vettel saw the checkered flag indicating that he had won the inaugural Abu Dhabi Grand Prix in 2009, it is very unlikely that he was aware of the eventful period leading up to the racetrack's construction. The Yas Marina Circuit had just been finished in record time to host the final race of the season. Commissioned by Aldar Properties PJSC, one of the largest developers in the United Arab Emirates,

Figure 1–5 Arup, Aldar HQ Designer's impression and detailed construction model including steel and concrete detailing.
Courtesy of Arup

the racetrack is one of a number of major architectural and urban projects built on Yas Island, just off the Abu Dhabi coast. The racetrack is adjacent to other architectural marvels, planned previous to the financial crisis of 2008, such as the Ferrari World and the Abu Dhabi National Exhibition Centre, which conspicuously display the wealth of this region that is rich in natural oil resources.

Abu Dhabi leaders had high ambitions to present their country and their culture to the world as a modern society. Aldar Properties PJSC wanted to be part of that effort when it came to the development of their own headquarters in 2007. They commissioned an iconic landmark building, an architectural and engineering masterpiece, to grace the Abu Dhabi Skyline.

What followed was the appointment of some of the world's leading experts in design, engineering, and construction in order to facilitate the fast-tracked delivery of the project. Lebanese architecture firm MZ Architects conceived a spectacular, semi-spherical (coin-shaped) concept for the 23-story building—the first of its kind in the world. The shape allowed for increased repetition in the facade panels, but at the same time posed challenges to the engineers and contractors as an entirely novel solution had to be found for the detailing and erection of the structure. The nature of the iconic form made steel the dominant construction material, as it was able to accommodate the high-tensile stresses inherent to the spherical shape.

The richness of fossil fuels in Abu Dhabi is not matched by an equal richness in high-grade steel. It cannot be sourced locally in the United Arab Emirates. This became obvious to the Head Contractor (UK–based Laing O'Rourke) and the Engineering Consultant Arup early on in the project. A solution had to be found to procure high-grade steel from the United Kingdom and to orchestrate the entire design, engineering, logistic, and construction process around a unique method for supply chain integration. With production of steel one issue and transport another, the tight time frame of a DnC (called D-B in North America) procurement pushed Laing O'Rourke to look for new ways to make the link between design, engineering, fabrication, and construction. Given the lack of a contractual precedence that would specifically address the issues apparent in the project, Laing O'Rourke put forward a teaming agreement to manage the collaborative workflow of the team using BIM authoring tools. They found a strong partner in Arup's Sydney-based "Regional BIM Coordinator" Stuart Bull and Dubai-based Steve Pennell, who were prepared to shoulder some of the risk of entering unchartered territory. Bull, who is now Managing Director of Ridley VDC, had previously been engaged as a virtual construction integrator on a range of high-profile construction projects globally, ranging from Foster + Partners' City Hall (2002) to PTW, Arup, CSCEC, and CCDI's Beijing National Swimming Centre—the Water cube (2008). Bull knew that the only way to meet the client's tight schedule was to collaborate closely with Laing O'Rourke and the UK steel fabricator William Hare (WHL) in order to produce a virtual shop model that could translate directly into constructible elements. The concept of the virtual model had to be aligned with UK and Abu Dhabi codes and regulations as well as the supply chain integration of various suppliers and forwarding agents. A global collaboration ensued, in order to facilitate just-in-time construction with information being shared between Australia, the United Kingdom, and the Middle East. In reflecting on the key point of difference that allowed for the team to succeed, Bull recalls a week-long design meeting of WHL key project staff in Sydney in order to resolve steel grade, material availability, and fabrication issues on the virtual model collaboratively. The teaming agreement helped to facilitate a highly collaborative and outcome-focused process among stakeholders; it was one of the key factors informing its success, enabling the facilitation of BIM use on the Aldar headquarters.

There were lessons to be learned. In the absence of a dedicated BIM Manager on the project, a Laing O'Rourke Project Manager had to step into that role. Further, there had been no dedicated BIM Execution Plan available on the project and many aspects of the collaboration had to be tested in the heat of the fast-paced project delivery. One of the biggest regrets of the team was the lack of integration of their highly sophisticated virtual construction model with the Building Maintenance Contractor.

The above example highlights that BIM doesn't work (well) when a predominant focus on gaining advantages gets applied by individual stakeholders. Despite the risks inherent to multidisciplinary collaboration, leading examples highlight the benefits of increased collaboration and sharing among project stakeholders, based on trust and respect among collaborating consultants and contractors.

The Aldar HQ example refers to one particular case for a high-profile project, undertaken in collaboration over three continents by highly experienced operators. It was one of the earlier examples where supply

Figure 1–6 ALDAR Headquarters detailed construction model including steel and concrete detailing by Arup.

Courtesy of Arup

chain integration via BIM between consultants and contractors led to a successful outcome. BIM has kept evolving since the completion of the Aldar HQ project, around the time of the 2009 Abu Dhabi Grand Prix. Numerous industry studies illustrate the steady increase in global BIM adoption.[13] According to an annual report issued by the UK National Building Specification (NBS), BIM use there rose from 39 percent of respondents in 2011 to 54 percent in 2014. More dramatically, a Smart Market report issued by the U.S. publisher McGraw-Hill in 2012 illustrates an increase in levels of BIM adoption in North America from 28 percent in 2007 to 71 percent in 2012.[14]

With increasing adoption we also start to understand where we struggle to get BIM right. The following section points out the most common examples of BIM going wrong. It is based on feedback from those who sit in the trenches and who deal with the consequences of badly implemented BIM day by day.

When BIM Goes Wrong—Examples of "Bad BIM"

MAKE MISTAKES FASTER

ANDREW GROVE, CO-FOUNDER, INTEL

What can we learn from BIM gone wrong? What are the key mistakes that repeatedly seem to creep into our projects and that sideline our best intentions when applying BIM in practice?

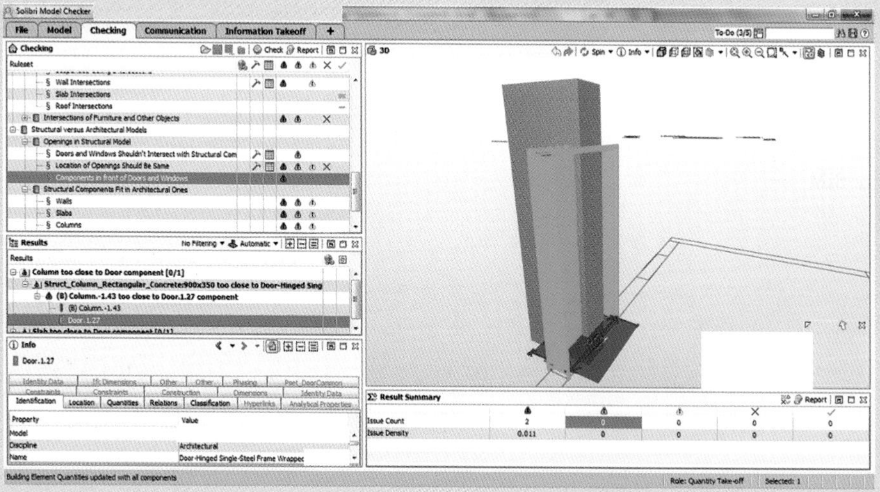

Figure 1–7 Detecting coordination issues in BIM via a model checker by Mitchell Brandtman 5D Quality Surveyors.

© Mitchell Brandtman 5D Quality Surveyors

It is arguable if "Bad BIM" exists. Let's assume for a moment that the core concepts behind BIM are noble and that they aim at improving how projects get delivered across a building or project lifecycle. Following that thought, BIM cannot be bad as such. It certainly can be interpreted, or applied in a bad way. Its principles can be overlooked, and its goals can be misunderstood or misused.

Anyone who ever worked on a project using BIM will have a number of stories to tell about BIM going sour. In some cases, the apparent shortcomings may have little to do with BIM itself, and they rather depend on the specific project context (for instance, contractual constraints or procurement problems). In others, the shortcomings could refer to a lack of skill or knowledge about how to get the most out of BIM. A good portion of "BIM going wrong" can be attested to the fact that those implementing BIM are often still going through a major learning curve across project teams.

Some common mistakes stand out. It is crucial for any BIM Manager to learn from those mistakes in an attempt to avoid falling into the same. The following hit-list represents a summary of responses from 40 of the world's leading BIM Managers[15] who operate across the United States, Europe, Asia, and Australia.

Pseudo BIM

It may sound implausible at first, but the biggest challenge faced by BIM Managers is what best can be described as "Pseudo BIM." There exists a spectrum of BIM "pseudoness" eventuating in practice. In its worst form, pseudo BIM is used to pretend BIM was applied whereas in reality a traditional CAD workflow was used to deliver a project. The reasons for such deception may be to impress clients (who may not know the difference), or to conform to client/regulatory requirements. Array Architects' Robert Mencarini describes this occurrence as follows: *When some team members think that working in CAD and then creating a model at the end of a phase constitutes true BIM, this isn't BIM and it creates problems.* Instances of this form of dressing up and masquerading are on the decline as more and more clients and/or authorities become more informed about the distinction between Pseudo BIM and the rest.

The most common occurrence of Pseudo BIM is applied by those who use BIM tools simply to produce their 2D documentation. BIM software gets utilized as a means to generate submission documents more efficiently. Multi-disciplinary coordination or data-integration opportunities are not considered by teams who separate geometry from data. The crime committed relates less to any active act of deception, but rather to the cowardice of going out of one's comfort zone—the 2D CAD workflow. The negative effects on other project team members are severe. This form of Pseudo BIM shuts the doors on any form of information sharing beyond simple visual referencing. On a project level this usually plays out as a delay when single seemingly BIM-enabled project partners cannot commit to a BIM workflow. Others have to pick up the slack and gaps in the otherwise integrated approach of project delivery emerge. A common "subcategory" of this form of Pseudo BIM is the "Fall-back." Chris Houghton, Peddle Thorpe's BIM Manager in Melbourne refers to it as: *Hybrid BIM. Too much CAD. Either across the entire project team, or within a single business.* It gets applied by those who commit to using BIM, but who revert to 2D CAD part way through a project. There may be a number of reasons for that to occur. The most likely out of those is a lack of skill or support infrastructure to sustain continuing a BIM approach. It is likely the project leaders who pull the plug on BIM when they lose confidence that imminent submission deadlines can be met.

Considering the level of progress made by most organizations who implement BIM, it is surprising to see the Pseudo BIM Phantom still plaguing the industry. With time these issues are likely going to subside.

Going Solo—Lack of Coordination across Key BIM Stakeholders

One of the most difficult aspects of pitching a BIM approach to any firm is the fact that organizations tend to search for immediate benefits for their own business or members. It makes sense: Their key purpose is to make a profit or serve their members' interests. Therein lies a fundamental problem: Acknowledging that BIM can increase efficiency intraorganizationally, synergies only really kick in when applied across as a number of stakeholders involved on delivering projects. This circumstance needs to be experienced to be understood. In a project-based environment such as the construction industry, it is difficult for an organization to prioritize multi-disciplinary collaboration over its direct returns on investment. What is good for the project doesn't necessarily appear to be as good for the business.

BIM Managers are not the only ones who are caught up in this conflict, but they are the ones who most directly experience the tension between what would technically be possible and what appears wise from a business perspective. The more acquainted an organization is in delivering projects using BIM, the more likely they will acknowledge the need for collaboration. It is often the more experienced tier 1 or tier 2 design, engineering, or construction firms who push for collaborative BIM, by nature of the scale of projects in which they engage.

Problems emerge when modeling is done by different parties without sharing and overlaying these models for coordination. BIM is not used concurrently during the design phase, and design intent BIMs don't get made available to contractors as a reference during the fabrication and construction phases. Even further, it is still early days for consultant/contractor teams to consider the information needs of Facility Managers. Their work usually starts upon handover of completed facilities and their information requirements are different to those of architects, engineers, and contractors engaged in the design and construction phases. An uncoordinated solo BIM effort results in duplication of information in multiple, often barely interoperable, formats. Potential synergies are not being tapped into and the BIM process becomes inefficient when seen in a holistic project context.

The roots of the problems described above are manifold: Architects fear for their intellectual property and they are concerned about their professional liabilities, engineers see little point in generating models when the design isn't yet completely resolved (as they run the risk of having to accommodate costly changes constantly). Ill-informed contractors may feel inclined to discard design intent BIMs if they don't realize how they add value to their process. Facility Managers don't engage as they are either not on board with the project team yet, or they don't understand how BIM can assist achieving their objectives.

BIM Execution Plan—Lack or Lack of Use

A managed approach to execute BIM goes hand in hand with a focus on collaboration. Many problems related to uncoordinated collaboration emerge if teams don't develop and sign off on what is commonly known as BIM Execution Plans (BEPs), BIM Management Plans, or the like as early in a project as possible. Publicly accessible

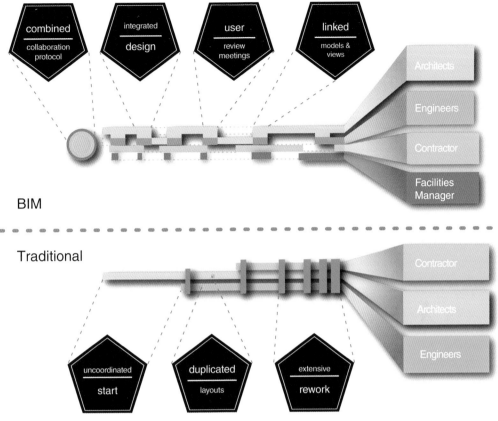

Figure 1–8 Comparing BIM versus traditional methods of delivery: Applying a combined protocol to regulate collaboration among stakeholders.

© Dominik Holzer/AEC Connect

BEPs templates have been available since 2007–2008 and these BEPs help orchestrate the entire collaborative process when using BIM. Their purpose is (among other things) to allow project teams to agree on the way models and associated information gets shared, how these models are put together, how often they are reviewed and who is responsible for advancing certain components inherent in a multidisciplinary BIM process. Their use—despite the fact that it is increasing—has not yet become standard on all medium- to large-scale construction projects. The absence of a BEP on these types of projects can lead to dissonances among collaborators and a loss in productivity. BEPs are by no means a guarantee for seamless collaboration using BIM, but they increase the chances for teams to work synergistically on declared and common BIM goals. Bad BIM happens when BEPs are either not available, or if they are not understood and adhered to.

No Data Integration

The next aspect of "Bad BIM" refers to an over-focus on geometric modeling to the detriment of associating data to the geometry that is useful downstream. With much attention given to generating 2D design documentation from 3D BIM, questions related to data often remain under-resolved. There is a good reason for that: Traditionally, consultants and contractors are rarely paid for adding any information to their documentation that doesn't directly benefit them. Convincing any team otherwise can be a difficult task if the inclusion of data requires an extra effort that is not covered by their fees. A recent study undertaken highlights the perceived extra effort by consultants of appropriating information from their models that may become useful by downstream parties.

Worse than the lack of data association is the process of developing datasets in parallel to existing BIMs. The two remain disconnected and (often incorrect) information gets mismanaged and doubled up in separate systems with incompatible formats.

Lack of Well-Defined Objectives (Client)

Those using BIM for design, engineering, and construction purposes are likely to be the key culprits for inconsistent approaches to BIM use. Still, clients, Project Managers, and Facility Managers are to blame for BIM going wrong as well. The "mother" of all aspects related to "Bad BIM" may be the lack of clear BIM objectives by clients. Such lack usually originates from an indifferent or uneducated client when it comes to declaring their information requirements at the outset, or at any point later in the project setup. Badly defined BIM objectives from the client side are often the cause for the lack of data integration described earlier. GHD's Brian Renehan laments clients who *over-specify goals, without an understanding of how the data will be generated, managed, and used.* Without declared and realistic BIM objectives, project teams usually tap in the dark as they need to second-guess what the client may be after. Business-savvy consultants and contractors see the opportunities of educating their clients and offering them help to uncover what data they may need at project handover. Others may not even have heard of support documents such as the UK PAS 1192:2 *Employer Information Requirements* (EIR) template, or the Asset Information Model (AIM) that gets created based on a Project Information Model that draws from the EIRs. Bottomline, lifecycle BIM cannot really work without an educated client who can articulate information requirements to the project team. The team may still develop data-rich models, but their usefulness is likely to be limited. The preceding BIM efforts may prove to be useless if clients don't specify what they want to get out of the project. Some clients try to play it safe by asking for "full BIM" or "fully integrated BIM" without the slightest idea how such elusive deliverables may benefit them.

Overmodeling

Stepping back from the client side, there exists another jewel in the crown of "Bad BIM"—it mostly occurs in the interaction between consultants and contractors (but it may extend to the world of Facility Management): Overmodeling.

"Over" may be a misnomer as it only reflects the most common occurrence related to the lack of understanding by a number of BIM stakeholders who are not in tune with the information requirements of their closest collaborators. NBBJ's Sean Burke describes it this way: *Concentrating on making perfect models—at the expense of useful and accurate data.* BIM gone wrong signifies in this context that there is too much (mainly geometric) information embedded in the model. Not only does this represent an unnecessary effort, it may well also make models too heavy to use, thereby jeopardizing the coordination effort across the team. The key reason behind overmodeling is insufficient communication between stakeholders. It becomes particularly problematic in the earlier design phases where an overload of information challenges the flexible design process. As much as a lack of data integration is bad, excessive modeling for the sake of including information (without a clear understanding what the information is good for) is just as problematic. Consultants (e.g., Mechanical Engineers) may go beyond suggesting systems and start to model detailed equipment only for the model to be turned over once the Mechanical Contractors get on board. Recently published guidelines about Levels of Development (LoDs) assist teams in harmonizing their modeling efforts. The definition of LoDs usually forms part of the BIM Execution Plans.

Lacking Tool Ecology

It is tempting to blame software vendors as the major culprits causing this problem. How often have they promised the world about the capacity of their tools? In fact, there is truth to claims that functions within the tools we use become more encompassing. By nature, software developers enhance their products over time in order to gain or maintain market share. When it comes to BIM tools, software vendors are quick to highlight their capacity to serve as sketch-design/conceptual modeling tools as much as they are suited for producing documentation output, 3D visuals, and data export to Facility Management. There may be truth to that, but questions emerge in how far "all-rounders" are the best fit for resolving any design/engineering/coordination/and data integration functions.

Problems emerge when BIM authors try too hard to resolve all design aspects with one model and one single software platform. They may be fixated on one way of doing things (just because it can be) without evaluating the best way of doing it.

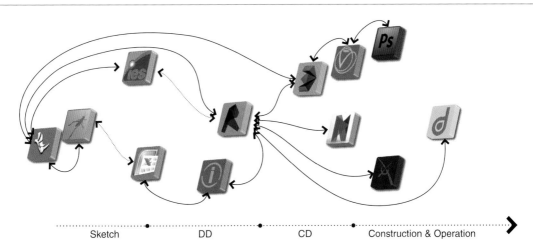

Figure 1–9 Strategic mapping of software interfaces to form a tool ecology associated with BIM delivery and beyond.

© Dominik Holzer/AEC Connect

This is where experienced BIM Managers step in. They know when and how to embark on the most appropriate pathway of connecting tools and passing on information in order to fulfill specific tasks. They understand how to establish a tool ecology and fine-tune data and geometry handover processes in order to maximize synergies within given suites of tools or across platforms.

Modeling without Understanding

More accounts of BIM gone badly refer to modeling efforts where those who author BIMs do not seem to be aware of the consequences of their proposed solution. As much as consultants are usually tasked with the production of "design intent BIM," contractors take over for the production of shop models for manufacture, detailed coordination, and installation. The nexus between intended artifact and the closest possible virtual representation of a construction component can be tricky at times. A spatially resolved model that is well coordinated and clash free is still no guarantee for success. It requires knowledge about constructability and serviceability in order to get BIM on an LOD 400 right. Such an understanding eludes most consultants (and even some contractors) and it comes with extensive site and shop-detailing experience. BIM with best intentions will still not work if those who model don't know how the project will ultimately be built and how certain components will be accessed for servicing and maintenance.

Model Inaccuracy

Parallel to a lack of understanding about what to model are problems related to a lack of knowledge regarding the requited model geometry accuracy. The accuracy associated with the generation of various BIM components and assemblies relates to the project phase and the construction material in question. Design intent BIMs tend to be delivered with lower accuracy than construction BIMs as consultants cannot be expected to know about precise construction tolerances by the various trades involved. Those trades are still liable for correct set-out and dimensioning of the virtual components that ultimately represent the equipment that goes up onsite.

Über-Hacks

One of the first words any BIM Manager learns on the job is "workaround": A way to achieve specific goals in BIM authorship and documentation outside the standard suggestion by the BIM software used. Workarounds are the bread and butter of BIM Managers. There exists a flood of webpages in support of workarounds. A culture of peer-to-peer support and communication has developed related to their use. In principle, workarounds can be seen as a positive option expanding the limits of any given software's tool infrastructure. In many cases, software developers learn from workarounds applied by the users of their tools and they may choose to integrate elements of those workarounds into future releases of their products.

Workarounds fall over when they become too complicated, or when they result in convoluted solutions that only benefit single authors. They may not be scalable across a team and even if one party benefits from a quick fix, others down the supply chain suffer the consequence.

Luckily, the richness of examples about unsatisfactory implementation of BIM can easily be matched by positive experience from practice. Respondents who provided their hit-list of "Bad BIM" examples were keen to share their views of successful BIM implementation. What should we aspire to deliver when working in BIM? What approaches to collaboration and project delivery promise increased efficiencies and synergies via the use of BIM? How can we ensure the penny drops and reap the benefits of a BIM workflow?

The Tipping Point—How Do You Become Successful Using BIM?

When the first settlers arrived in Sydney, Australia, to populate the penal colony in the late 1780s, they worked hard to establish a foreshore that would protect them from the prevailing tides and other elements. Back then as today, the sandstone coast of Sydney Cove shifts from rugged bushland to sandy beaches, jumping at times to form steep cliff-edges. Since the days of the early settlers, the coastline framing parts of inner Sydney has undergone a number of transformations. By the mid-1820s the first wharf was built at Walsh Bay followed by the wharfs of Millers Point.[16] Long wharfs that served for docking of trade and transport vessels have for decades been the most prevailing architectural/landscaping feature of an area that is now known as Darling Harbour—and more precisely, Barangaroo. One aspect of these wharfs is the lack of engagement they allow for inhabitants with the water. When the New York–based firm Johnson Pilton Walker in association with Peter Walker and Partners Landscape Architecture won the competition to design the new Barangaroo waterfront in 2009 they knew they wanted to address this issue. Their design surrounding the Barangaroo Headland Park represents the major urban redevelopment program in Sydney of the past 20 years. The team's rugged sandstone topography took inspiration from the naturalistic pre-1836 shoreline in order to allow the public to re-engage with the shore that has been locked away from them for more than 100 years.

The design for the new foreshore by the architects was simple and ingenious: sourcing sandstone found onsite in order to generate a differentiated series of blocks that step down toward the waterline. The arrangement of the 10,000 unique blocks is set in a way to allow the public to navigate different levels of the shoreline while being able to trace and engage with tidal variations in a lifelike fashion. Still, the arrangement of these tidal rock pools could not be arbitrary. Dimensioning, cutting, and transport of individual sandstone blocks, the overarching topography of the terrain, and height-limitations related to pedestrian circulation for easy navigation all formed constraints that needed to be addressed as part of the design. The team was stuck at a point where the ideals of the designers could not easily be broken down into feasible construction components by the contractor. The gap between design aspirations, engineering capabilities, local construction constraints, and cost factors had to be overcome.

As John Hainsworth, BIM Leader at Aurecon, explains: A tipping point was reached with the realization that the definition of the foreshore required a parametric approach to be taken with the input from the designers, the engineers, the contractor, and the stonemason. Not only was it important to rationalize the geometrical aspects of the design, but also the programming of stone-cutting, the QR-coding of the blocks, and the transport and positioning onsite. The geometric concept behind positioning the stepping sandstone blocks by the architects was well defined, but

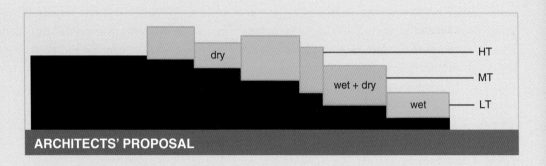

ARCHITECTS' PROPOSAL

dry

wet + dry

wet

HT

MT

LT

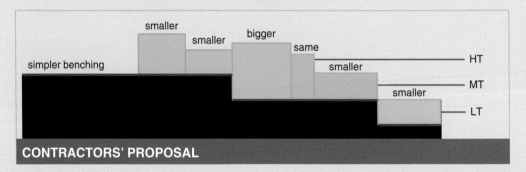

CONTRACTORS' PROPOSAL

simpler benching

smaller

smaller

bigger

same

smaller

smaller

HT

MT

LT

Figure 1–10 Aurecon, Barangaroo Headland Park Foreshore. Section comparing architect's and contractor's proposal for stone block arrangement.

© Aurecon

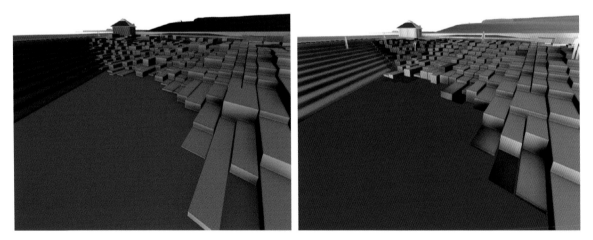

Figure 1–11 Aurecon, Barangaroo Headland Park Foreshore. 3D View comparing architect's and contractor's proposal for stone block arrangement.

© Aurecon

the head contractor—Lend Lease—had to overlay it with their own logic in order to make production and positioning feasible. Lend Lease, a great supporter of BIM-related processes, tasked the engineering consultant Aurecon to develop a data-driven BIM approach to resolve the issues mentioned above. Aurecon's model contained a rationalized benching beneath the blocks, but demonstrated a similar upper surface to match the architects' aspiration. The reinterpretation of the desired effect via integrated BIM processes resulted in a solution that was signed off by all major collaborating parties, which led to quick approval. Such success was in no way certain at the outset of Aurecon's involvement. The penny dropped for the team when Lend Lease pushed for a team approach, gathering key parties around a 3D model of the design as often as possible in order to resolve issues collaboratively. That way the team communicated and learned to understand methods for cutting the stone, the associated treatment of waste, and the mechanical treatment of the stone's surface. Buy-in by the client and well-orchestrated supply chain and fabrication integration via BIM by the entire team was the key factor for success.

The difficulties of introducing novel approaches to traditional contexts are well documented and described by Malcolm Gladwell in his book: *The Tipping Point—How Little Things Can Make a Big Difference*. There he concludes by encouraging those who are agents for change to *focus, test, and believe*.[17]

BIM—Getting It Right

How would those who count among the leading BIM Managers explain how to do it right? The responses that the author has solicited as part of the research for this publication from over 40 Design Technologists and BIM leaders draw a clear picture: If you want to do BIM right you need to think about the client first.

The number-one aspect of Best Practice BIM is to offer clients a better product and more certainty around the final outcome of their projects. "Certainty" as used here refers to a number of things.

First, BIM delivers clients a better understanding about their project through increased visualization opportunities. BIM also strengthens the design team's abilities to include environmental sustainability concerns early in a project's development. Tighter cost control about the planning and construction process can be applied when using BIM processes in the field. In addition, BIM allows the introduction of more transparency for construction scheduling and sequencing. Ultimately, those using BIM can pass on information from the construction to the operation phase of a facility. These are merely a number of aspects relating to the increased certainties that can be offered to clients via BIM.

The strong focus in client benefits expressed by BIM-enabled consultants and contractors may surprise at first. Those operating in the BIM space rarely represent clients. Still it makes sense if considered as part of the construction industry's push to establish BIM within a lifecycle approach. It also means that Best Practice BIM doesn't work in isolation, but that it requires collaboration across a project team. Next to client satisfaction, seamless design and construction coordination between consultants and contractors is of most relevance to leading BIM Managers. The industry is learning to adopt new pathways to make BIM work not just for companies in isolation, but increasingly also across the consultant/contractor divide. BIM gets used more and more to facilitate construction processes onsite and Field BIM as well as 4D programming are becoming an ever more relevant factor of good (or even best) practice.

Figure 1–12 Barangaroo
Headland Park Foreshore,
cutting stone blocks from the
onsite extraction hole.

© Troy Stratti

Figure 1–13 Barangaroo
Headland Park Foreshore,
stone blocks in their final
position.

© Troy Stratti

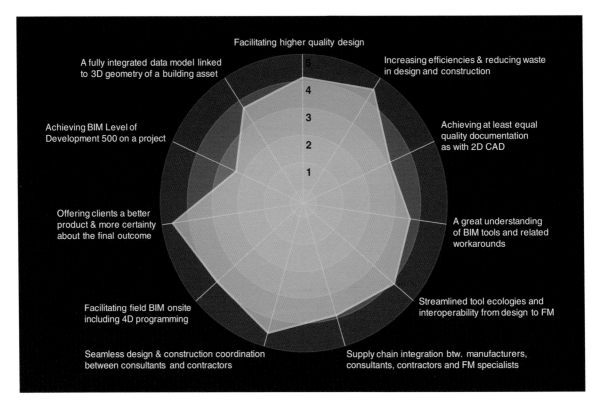

Figure 1–14 Responses from industry experts about what constitutes Best Practice BIM.
© **Dominik Holzer/AEC Connect**

Global BIM leaders are convinced that increasing efficiencies and reducing waste in design and construction is an absolute priority when considering BIM's best practice. Such a response needs to be seen in the light of the decrease in productivity across construction industries over the past 50 years in some western countries. Compared to other nonfarming industries, the dispersed nature of construction is highly inefficient with the doubling up of work and uncoordinated delivery approaches.

BIM experts highlight the need to streamline tool ecologies and to achieve interoperability from design all the way to Facility Management. A traditional project-delivery mindset usually doesn't consider supply chain integration and the alignment of tool infrastructures to facilitate information transfer from conceptual design all the way to operations. Respondents saw overwhelming benefit in BIM's potential for supply chain integration between manufacturers, consultants, contractors, and FM. Respondents supported the idea of a fully integrated data model linked to 3D geometry of a building asset and they highlighted the need for a BIM Manager's understanding of BIM tools and related workarounds. BIM Managers didn't believe that achieving at least equal-quality graphic output as with 2D CAD delivery was a high priority of BIM.

Those who assume that a sound knowledge of technical- and design-related aspects of BIM provide certainty for success in implementing BIM, should think again. As described in the foreshore example, getting BIM right depends on a different set of criteria. The primary driver to make BIM work is to ensure engagement and support of the upper management within an organization or on a project. It often entails changing the mindset of a firm's or a client's leadership in order to make them understand that BIM is more than simply a tool for delivering projects in 3D. Having full support from the top is a key enabler to roll out BIM in a sustained and structured way. If leadership is unaware, not involved, or doubtful about a BIM strategy, decisions get delayed and the implementation effort can easily get bogged down by micromanagement of secondary issues without a clear plan or direction. Another prerequisite for success for succinct BIM implementation is the attitude by the team when it comes to project delivery. The more the team embraces a BIM workflow, and the better they communicate their requirements, the more likely BIM will provide them with tangible benefits. There is no surprise in such a statement. Still, a good number of teams underestimate the value of adhering to a well-conceived BIM Execution Plan in order to tap into the full potential of what BIM has to offer. Understanding BIM as a team sport and adhering to guidelines that were defined in collaboration doesn't come natural to some organizations. It requires a maturing process where—at times—firms put the advantages of the team ahead of their own. Feedback from industry experts suggests that such a maturing process and the implementation of Best Practice BIM typically takes an organization three to four years or more to master.

One of the reasons for this extended adoption period is the lack of clear directives, or "pull" from the client side. Firms push in a direction without necessarily being fully aware of the BIM end-goals by their clients. Clients therefore play a crucial role in establishing overarching BIM goals on projects. By defining what those goals are, clients (while still considering their own benefits) provide teams with an orientation point to work toward.

Benchmarking BIM

What metrics can one apply to measure the quality of BIM? What are the Key Performance Indicators associated with such metrics?

Broader Policies

The quantitative capture of BIM performance has been up for debate for a number of years. On a policy and an industry level, a number of governments or industry bodies have come up with their own breakdown of BIM into defined levels or stages. In some cases (such as with the UK PAS 1192) the lifecycle aspect of BIM is given high priority; in other cases benchmarks are scaled down to more immediate and targeted aspects of BIM that serve to satisfy a department's needs, such as Spatial Programming by the U.S. General Services Administration (GSA) or the Submission of BIM for planning approval/permitting processes by the Building & Construction Authority (BCA) in Singapore. There, the BCA stages the requirements for mandatory BIM e-submissions for architectural and engineering approvals for all new building projects of a certain size.

These guidelines often provide overarching frameworks for BIM Managers and their teams to steer their efforts toward a certain direction. They are an orientation point—a beacon for guiding industries toward higher efficiency in the use of BIM. But there is much more that needs to be considered. For BIM Managers, the key deliverables correspond to their output on a project level. What is it that BIM Managers are most concerned with in everyday practice? How do they measure their success—or lack thereof? The metrics and benchmarks presented here stem directly from feedback given by the 40+ BIM managers who took part in the study leading up to this publication.

Measuring Day-to-Day Performance

The results highlight one crucial factor: A singular formula for Best Practice BIM doesn't exist. Top benchmarks for Best Practice BIM vary from stakeholder to stakeholder and even from project to project. A number of general trends and tendencies still cut across an otherwise diverse set of criteria.

If we believe the feedback from the experts, the most relevant metric for successful BIM is outward looking: Client satisfaction! As Dennis Rodriguez—BIM Enterprise Manager at the global engineering firm AECOM—puts it: *A fully integrated data model attributed for the client's use for facility management and operation is essential to the market realization of the true value proposition of BIM.* A key benchmark therefore relates to the quality of data that can be generated via BIM and made available for clients' FM purposes. Toby Maple, National BIM

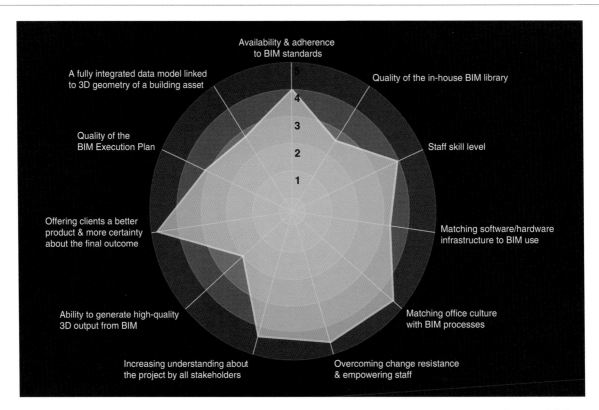

Figure 1–15 Responses from industry experts about the metrics applying to Best Practice BIM.
© **Dominik Holzer/AEC Connect**

Manager at Australia's largest architectural practice HASSELL, adds that an important prerequisite to facilitate such handover is the project team's ability to first *articulate the "value" to various stakeholders, whether that is for the client, consultant, builder, owner, FM expert, or others.* Based on a number of pilot studies undertaken in the United Kingdom, Mark Bew (Chairman of Building Smart (UK) and Chairman of the UK Government BIM Group) sums up key benchmarks for Best Practice BIM as: *dramatic fiscal and quality improvements.* This promise of BIM aims directly to keep projects on time and on budget. How do teams achieve cost savings while increasing output quality?

Expert BIM Managers agree that the quality of documentation and the smooth delivery of projects in collaboration is a crucial driving factor behind BIM. Improvements in that area can be measured via the reduction of coordination issues and Requests for Information (RFI)s or Change Order Requests. Further benchmarks are inherent to a reduction of waste by avoiding single-use model generation by sharing coordinated models that add value to multiple stakeholders' activities. Casey Rutland from Arup Associates in London points out factors that allow for such coordinated and targeted BIM work to unfold: *Contractual and project management documents agreed and used through appointment.* The adherence to well-configured BIM guidelines such as an Employer Information Requirement document of a BIM Execution Plan represents another benchmark for

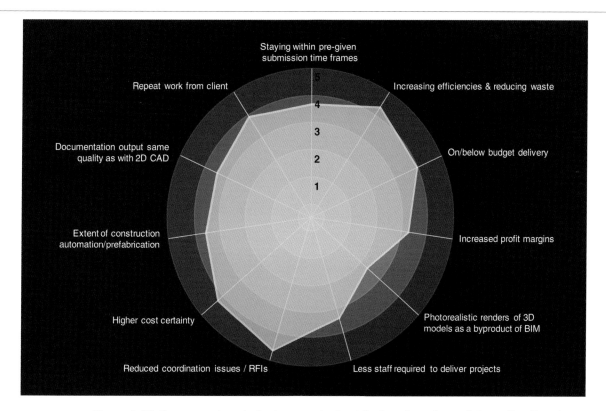

Figure 1–16 Responses from industry experts about the benchmarks applying to Best Practice BIM.

© Dominik Holzer/AEC Connect

Best Practice BIM. Hand in hand with the reduction of waste comes higher cost certainty and a reduction of risk. Adam Shearer, BIM expert at YTL in Kuala Lumpur (Malaysia), touches on the risk issue by stating: *Best Practice BIM is about using innovative technologies to predetermine and lower risk to the benefit of the stakeholders and the client.*

Other metrics for Best Practice BIM relate to inner-organizational benchmarks, namely the cultural context within a firm. Overcoming change resistance and empowering staff is seen by experienced BIM Managers as the primary goal within their organization. This revelation points toward a crucial cultural aspect related to BIM: The high importance of Change Management in association with implementing BIM. Despite much attention given to the technical aspect of implementing BIM within and across organization(s), the cultural side tends to be neglected. Such is the relevance of Change Management, that part of this publication is dedicated entirely to the topic. Staff empowerment is a crucial factor of a well-considered Change Management strategy. BIM Managers are specialists and one of their key tasks is to convey and share a portion of their knowledge to others in order to empower them to fulfill their tasks better. How can one measure such knowledge transfer? How does a BIM Manager ensure the empowerment of others with the work he or she does? At times empowerment occurs as part of day-to-day mentoring provided by BIM Managers to others, at other times it is reflected—less directly—in the quality of "back of house" documents such as BIM standards, BIM Execution Plans, and more. Expert BIM Managers assign high priority to the availability of high-quality BIM standards. They are the ones accountable for establishing such standards and they need to ensure that staff adheres to them across an organization.

The diversity of issues listed here reflects the complexity BIM Managers are faced with. On one hand, they are tasked with helping achieve lifecycle goals on a project even though these may stand in conflict with the understanding by upper management of what's best for their organization. On the other hand, they need to have a great understanding of design and construction processes in a highly interactive environment. In addition, they need to be fluent in the use of a range of software applications and understand how to combine their use efficiently. On top of all of this, they need to be great communicators with great people-management and communication skills.

Asked about the most relevant tasks for BIM Managers, exerts report the following:

Overseeing BIM-related process and workflow ranks first, with the facilitation of multidisciplinary coordination and the development of BIM Execution Plans coming second. The third most relevant task for BIM Managers is the link between office/project leadership and BIM authoring tasks. It is essential for BIM Managers to determine standards for information and knowledge management and they need to be strongly involved in assisting their organization in making the right choices when employing new staff.

It is crucial for BIM Managers to grow a culture of support, instead of attempting to provide all the support themselves. When asked about this issue, expert BIM Managers ranked the "provision of assistance on the floor" lowest out of all possible answers. Within an environment that is mainly focused on design exploration and delivery, technology-related aspects often become secondary to some, in particular if all they want is immediate support without considering adding to their knowledge. BIM Managers easily get caught in this conundrum; their role is mistaken for project support which is not the same as managing BIM. Despite any expectation by

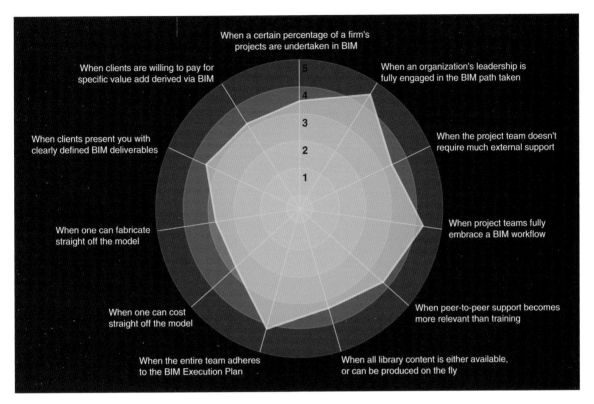

Figure 1–17 Responses from industry experts about the tipping point for achieving Best Practice BIM.

© Dominik Holzer/AEC Connect

others and/or eagerness by BIM Managers themselves to "help out," it is the BIM Manager who clearly has to establish and communicate his or her role beyond project support. Empowerment doesn't occur and efficiencies are not gained if BIM Managers keep their knowledge to themselves and get drawn too deeply into project work. This is a problem faced throughout the industry. To a degree it stems from miscommunication between BIM Managers and an organization's leadership. The need to engage an organization's leadership about BIM is clearly expressed by experts who highlight what they perceive as the tipping point for Best Practice BIM.

Key Performance Indicators

There exists a range of Key Performance Indicators (KPIs) for BIM Managers to consider. As much as BIM Managers can seldom measure client satisfaction, it sits within their reach to maximize the quality of their output while aiming for the highest possible efficiencies to get there.

When looking at KPIs for successful BIM, some see a proven track record of successful projects as the most relevant aspect. Once an organization has successfully delivered its first few projects using BIM, the BIM Manager is in a far better position to demonstrate performance based on tangible outputs such as drawing sets, 3D

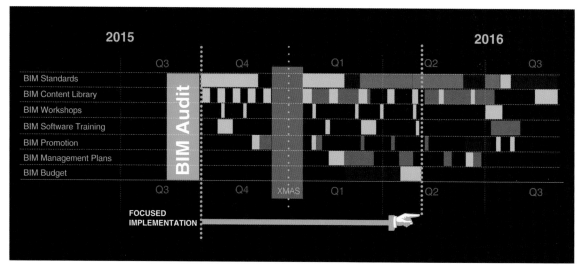

Figure 1–18 Mapping BIM KPIs against timelines for implementation.
© **Dominik Holzer/AEC Connect**

renderings, and data output. Showcasing the finished product after a period of fine-tuning the quality of BIM drawing sets works a thousand times better than trying to explain that the use of BIM may not lead to a decrease of documentation standards.

Another BIM KPI is the interface between the model and the 2D documentation output (or 4D/5D scheduling and costing when considering BIM for contractors). The best practice approach therefore is to ensure that model authors adhere to a clear set of BIM standards, which in return correspond to a well-configured, standardized, and lean BIM object library. If set up correctly, the representation of model information as 2D documentation can then be automated to a large degree via the use of well-structured view templates or filters. Similar arguments can be brought forward for the interface of models with coordination and programming software as well as quantity takeoff. This holy trinity of documenting in BIM—standards, library, and view templates—can be expanded to serve lifecycle benefits. The 2D output becomes a byproduct of increased data integration from specification to documentation, construction, commissioning, and operation and maintenance (O&M). BIM Managers need to use in-house BIM standards as a starting point for producing BIM Execution Plans that help regulate the multidisciplinary collaboration process. The quality of the template documents that feed into those is another KPI that sits in the BIM Manager's corner.

Any BIM standard, or well-structured library or view template, is only as useful as the systems in place to ensure relevant stakeholders adhere to them and apply them correctly in their day-to-day work. BIM Managers need to go through a constant process of monitoring and Quality Assurance (QA). Therefore, a further KPI for them is the level of reporting with their collaborators such as Model Managers or BIM coordinators. Regular model audits are as crucial as weekly meetings with key BIM stakeholders. Based on these meetings, BIM library content needs to be reviewed continuously and BIM standards should undergo regular revisions.

In the context of BIM, skill development KPIs can be assigned to the level of a BIM Manager's involvement in recruitment, the availability and quality of a BIM induction process for new additions, and the strategy for

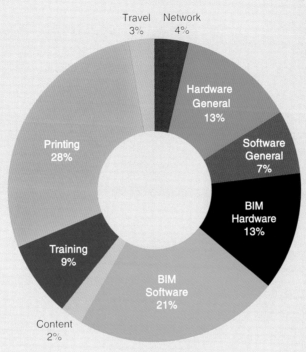

Figure 1–19 Establishing a Design Technology Budget with itemized listing of key cost factors.
© Dominik Holzer/AEC Connect

advancing a colleague's BIM skills to the desired level. In addition, BIM Managers are also responsible for the promotion of an organization's BIM capabilities both inside and outside the firm. Regular newsletters and in-house presentations are important. The generation of BIM Capability Statements for tenders and other forms of promotion are a must. No organization can afford to neglect the public's perception of their BIM efforts. In some cases, BIM becomes a prerequisite for winning work in the first place.

A KPI that sometimes gets overlooked is the Design Technology Budget. Does it exist? Does it separate between capital and operational expenditure? How can it be set up so to become a useful decision-support instrument for upper management? By itemizing and grouping various cost-related expenses associated with Design Technology and IT, a BIM Manager can start to demystify an organization's budget related to BIM. The Design Technology Budget thereby becomes a crucial ally in order for BIM Managers to establish business cases, justify current expenses, and plan ahead strategically.

There is one major set of KPIs that have remained unmentioned so far. They all relate to a BIM Manager's ability to guide an organization through change. Chapter 2 of this publication is entirely dedicated to the topic of Change Management. The underlying social, psychological, economical, and organizational effects related to the introduction of highly disruptive technology such as BIM will be explained.

Thank you to all the experts who so generously offered their thoughts and insights for this chapter:

Julia Allen of HASSELL, Aleks Baltovski, James Barrett of Turner Construction, Mark Bew of ECS Ltd, Cory Brugger of Morphosis, Stuart Bull, Sean Burke of NBBJ, Ronan Collins of InteliBuild, Mark Cronin of Peddle Thorp, Jon David of Turner Construction, Josh Emig of Perkins+Will, Gustav Fagerstrom, Bruce Gow of COX Architects, John Hainsworth of AURECON, Belinda Hodkinson, Chris Houghton of Peddle Thorp, Jason Howden of Warren and Mahoney, Dan Jürgens, Stephan Langella, Jan Leenknegt of BIG, Michelle Leonard of Assemble, Peter Liebsch, Toby Maple, Bruce McCallum of Dialog, Robert Mencarini of Array Architects, David Mitchell of Mitbrand, Chris Needham of AECOM, Paul Nunn of PDC Group, Chris Penn of Hansen Yuncken, Alexandra Pollock of FXFOWLE Architects, Brian Renehan of GHD, Rachel Riopel Wiley of HDR Inc., Dennis Rodriguez of AECOM, Bilal Succar of BIM Excellence, Casey Rutland of Arup Associates, Adam Sheather, Warwick Stannus of A.G. Coombs, Chris Tate of Architecture BVN, and Robert Yori of SOM.

Endnotes

1. D. Neeley, "The Speed of Change," CMD Group, June 14, 2010, http://www.cmdgroup.com/market-intelligence/articles/the-speed-of-change/ and "The Business Value of BIM for Construction in Major Global Markets," McGraw-Hill Construction, March 2014, http://analyticsstore.construction.com/GlobalBIMSMR14.

2. J. Laiserin, "Comparing Pommes and Naranjas," *The Laiserin Letter*, December 2, 2002, http://www.laiserin .com/features/issue15/feature01.php.

3. G.A. van Nederveen, and F. Tolman, "Modelling Multiple Views on Buildings," *Automation in Construction*, December 1992, Vol. 1, Number 3, pp. 215–224.

4. Object-oriented modeling systems, such as Oxsys, had their origins in the 1970s. P.H. Jurgensen, "The Conceptual Form of Architectural Design", Master Thesis, Massachusetts Institute of Technology (MIT), 1986, p. 19.

5. Course Details—BIM Management, BCA Academy, https://www.bca.gov.sg/academy/courses_tests.aspx? Course_Exam_Code=CS00151.6.

6. Membership Assessment Criteria and Rights, The Hong Kong Institute of Building Information Modelling, http://www.hkibim.org/?page_id=10.

7. Certificate of Management—Building Information Modeling, AGC of America, http://www.agc.org/cs/cm-bim.

8. UK BIM Level 2—B/555 Roadmap, A Report for the Government Construction Client Group, BSI (UK) March 2011, http://www.bimtaskgroup.org/wp-content/uploads/2012/03/BIS-BIM-strategy-Report.pdf.

9. BRE announces BIM Level 2 training and certification pathway, April 23, 2014, http://www.bre.co.uk/news/BRE-announces-BIM-Level-2-training-and-certification-pathway-969.html.

10. CanBIM Certification, Canada BIM Council, http://www.canbim.com/canbim-certification-0/what-is-canbim-certification-83.

11. *See also* buildingSmart Alliance, http://www.buildingsmart.org/.

12. "State of Ohio Building Information Modeling Protocol," Ohio DAS—General Services Division, 2011, http://ofcc.ohio.gov/Portals/0/Documents/MediaCtr/M830–01-BIMProtocol.pdf.

13. *Ibid*, D. Neeley, McGraw-Hill Construction.

14. NBS (UK), National BIM Protocol 2014, http://www.thenbs.com/images/bim/NBS-National-BIM-Report-2014_infographic.png.

15. In order to research for this publication, the author has approached approximately 40 high-profile BIM experts who are currently working in small to large design, engineering, construction, or other consultancy practices across several regions globally. The experts responded to a 16-point questionnaire and they shared their knowledge via in-depth insights from their practical experience.

16. Barangaroo History, http://www.barangaroo.com/discover-barangaroo/history.aspx.

17. M. Gladwell, *The Tipping Point—How Little Things Can Make a Big Difference*, Little, Brown and Company, 2000, p. 253.

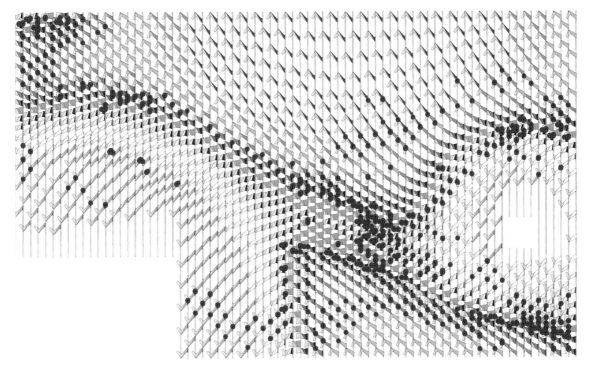

Figure 2–2 Emerson Los Angeles project by Morphosis: A geometric model that contains/embodies a large amount of information including constructability, costing, fabrication, and design.

© **Morphosis Architects**

between the team's goals and the legal framework by which they are bound. Within individual organizations tensions arise due to varying levels of engagement of staff with technology, the associated range of skills, and varying levels of resistance to accommodating change to well-established workflows. In addition, an organization's leadership is not always aware of the impact BIM has on its business.

The Cultural Dimension of Change . . . and Its Management

Organizational change requires a change in culture. Social Science describes organizational culture as: "the constellation of values, beliefs, assumptions, and expectations that underlie and organize the organization's behaviour as a group."[4] Cultural change therefore has a profound impact not only on an organization as a whole, but also on its individual members as it is asking them to alter their perception and behavior.

The Social and Organizational Context to Change

"An organization doesn't change until the individuals within it change."[5]

In the construction industry, there exists a tendency to approach a project with the same methods that have worked in the past; successful delivery often depends on preestablished "formulas" and workflows that have proven to be efficient and effective for those involved. Due to the traditionally poor knowledge-capture in the construction industry, many of these formulas sit within individual project leaders' heads.

The advent of BIM scrutinizes this approach. The introduction of high-end technology for information sharing requires a re-think of established workflows. BIM implies a more structured and less individual-centric approach to knowledge-capture. Project team constellations need to be revised in order to help manage the increased data connectivity across a project team. Project leaders are likely to be challenged by the increase in information management required to share models efficiently among various participating stakeholders.

What needs to happen to empower those who are well set in their traditional ways to open up toward new approaches? How can one structure a process of engagement that leads to sustained organizational change?

In his publication *Critical Success Factors of Change Management*,[6] Tim Fritzenschaft suggests three major phases associated with Change Management: Change agents first need to *prepare and create readiness for change*; before *executing change*; and subsequently *consolidating* it.

The first phase requires a clear definition of the objective and an analysis and understanding of the current situation or environment to be changed. It is pivotal to create a shared awareness among affected stakeholders about the problems that lead up to the need for change in the first place. In addition, change facilitators should have the ability to communicate well to those affected what the upcoming changes will encompass. What effect are they going to have on them personally? Creating a shared understanding about the areas of change required both, in terms of buy-in from upper management as well as operators in the trenches, will soothe the disruptive nature associated with change.

Fritzenschaft describes the second phase as the crucial point where change agents determine competences and responsibilities that lead to change. This is the phase where employees get involved in executing the change. At this point, it is crucial to identify key organizational roles related to the facilitation of change. It is also the period where most resources (time, money, and manpower) are required. This is the time of dialogue and engagement. Whereas the audit undertaken in the first phase results in awareness about what needs to happen and how to achieve it, phase two encompasses a period of transformation. This phase has a major impact on the cognitive landscape of those affected.

Any organizational change is only as effective as the means to uphold and sustain its effects. Therefore, the final phase, according to Fritzenschaft, is all about consolidation of what has been achieved. It is the period where results get communicated and progress gets monitored continuously.

BIM Managers: Facilitators of Change

BIM Managers play a decisive role as change facilitators. Change facilitators assist key stakeholders within an organization by mentoring them on their path to deal with change. They empower them to engage with a changing context that affects their professional and personal life. Their involvement is to determine a vision and a program for implementing the three phases as described above. First, BIM Managers need to base their work on a strong awareness of the situations they encounter within their organization and beyond. They are the ones who understand the broader industry context when it comes to technology uptake and its reconciliation with existing practices.

During the process of facilitating change, BIM Managers train, mentor, and build communities in order to align traditional workflows with new approaches that respond to the strategic BIM and technology direction identified earlier. This is the period where those affected are likely to be taken out of their comfort zone, the period where BIM Managers empower them to cope with and embrace change that affects them.

In order to consolidate their efforts, the BIM Manager needs to monitor the adherence of staff to particular workflows, standards, and other BIM-related processes that were introduced in the second phase. There exists a tangible danger for those who are new to adopting BIM to fall back into old habits once problems arise. It should be the declared goal of any BIM Manager to assist an organization to get to a point where the benefits of using BIM outweigh its upfront investment. Depending on the quality of the BIM Manager and his or her associates, this point happens sooner or later. A well-developed strategy will also assist the practice to decrease their effort levels in delivering projects altogether.

Change Management in the context of BIM is by nature an ongoing process. There is clearly a major upfront learning curve to master: Those at the start of embracing BIM are confronted with a large number of major

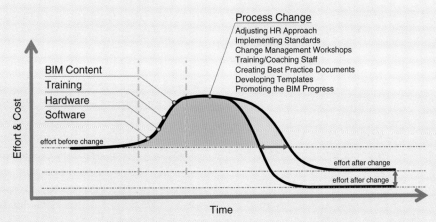

Figure 2–3 Change Management graph, reducing the effort of project delivery.
© Dominik Holzer/AEC Connect

changes to traditional workflows. At the same time, even those with a proven track record of successful implementation across their organizations (and with a number of reference projects to show for it) still highlight that they are on a path of discovery. BIM keeps evolving and it does so across a widespread range of building lifecycle–related topics. If architects and engineers were first to apply BIM to facilitate design, engineering, and documentation, contractors started soon thereafter to accommodate it for construction coordination and management. As time progresses, more and more clients/developers/project managers/and manufacturers are getting involved. Inherent to a rise in information sharing capability across an expanding group of stakeholders comes an increase in complexity and the need for Change Management. The gradual expansion of the BIM constituency has occurred over the past five to ten years and it is still ongoing. Hand in hand with the expansion of stakeholders is the availability of technology that allows them to interact via BIM. BIM Managers therefore deal with a moving target when helping to facilitate an organization's response to the implementation of BIM.

Interfacing with Your Organization's Leadership and Management

BIM as a process is highly disruptive to traditional practices applied across the construction industry. In addition, BIM continues to evolve at a rapid pace. Those new to its implementation often struggle when taking their first steps; those already proficient in its use need to ensure they keep up with the rapid changes as part of its development. In that sense, BIM has a major impact on business aspects of a large proportion of stakeholders across the AEC industry. Still, there appears to be a disconnect between those who manage and lead an organization and those who actively engage with BIM. It is the BIM Manager's role to bridge this gap. The reasons for this disconnect are manifold, but one of the key factors is that organizational change associated with BIM usually occurs tangentially: It neither emerges directly via bottom-up demand from staff (except possibly in a select few cases), nor does it typically result from a top-down directive. Instead BIM is often introduced by a technology specialist within consulting or contracting firms, or simply by an enthusiast who is passionately pushing for change. One other reason for an organization's leadership skepticism toward BIM is the fact that the associated lifecycle thinking at times stands at odds with the immediate business interests of an organization. Leaders tend to focus on internal benefits rather than what is good for the project. With the absence of a revised scheme for redistributing fees, leaders see the risk of their organizations taking on more work than they are paid for.

BIM's Push and Pull

When using BIM we do not just apply new software to replicate processes we used to engage in less efficiently in CAD. When using BIM, we drastically alter established workflows, relationships, and deliverables. We thereby impact a wide range of stakeholders across the construction industry and beyond. The kind of change required, and the extent to which change is needed to facilitate BIM, is often unknown to upper management. Due to their lack of awareness about BIM, they struggle to decide on what to change, which direction to take, or how to achieve the change process. The lack of client pressure or mandates by authorities results in the predicament that BIM-related organizational change gets driven from within the organization (push) rather than presenting a clearly defined target to work toward (pull). At the same time, managers tend to be highly alert to changes triggered by outside factors such as e-market demand, policies, or other business-related circumstances.

Richard Saxon predicts a change in market behavior with more pull from the client side: "BIM is a supply-side phenomenon, offering changed performance to the market as 'Push.' What will determine how it changes the industry will be the customer 'Pull,' the services actually sought by the market."[7]

The internal push for BIM remains problematic if upper management does not realize that its implementation is in essence a management-related task. Management often mistakes it as a technical challenge that is best resolved by IT staff, CAD experts, or junior staff members with a technology edge. Even if overarching policies, or even mandates, exist (there are increasing numbers of those emerging globally), they are more likely than not going to be read and understood by the BIM Managers and not by upper management. The consequences are problematic.

Decision Makers Who Do Not Understand BIM

Within those organizations that are new to the adoption of BIM, upper management's relationship with BIM Managers is at times rather reactive instead of proactively engaging. In these instances, it is common that the communication flow between a firm's decision makers, project leaders, and BIM Managers and authors is not well structured; it remains a one-way street. The purpose of BIM and its goals get communicated insufficiently, with little or no benchmarks for implementation and strategy for change. BIM's effects on business remain unexplored. Such shortcomings result in miscommunication between stakeholders and misunderstanding of the impact of BIM on projects. It can lead to annoyance by project leaders who do not seem to get what they want and it simultaneously can lead to frustration among BIM Managers who feel misunderstood or disrespected. Along with the lack of understanding of BIM itself, the role of the BIM Manager as an agent for change is, at times, little understood or supported by upper management.

Based on feedback from expert BIM Managers globally, the lack of engagement from upper management and/ or the client represents the primary obstacle BIM Managers need to overcome. The most knowledgeable BIM Manager who is surrounded by the most competent team will still struggle and likely fail to establish best practice BIM if upper management does not understand it in the first place. Such lack of engagement is reflected in a number of typical examples:

- Obscured views about what BIM is

- Mistaking BIM for 3D CAD

- A mere focus on software and technology

- A lack of understanding of how BIM impacts HR or staffing on projects

- Recruiting staff that lack the required BIM skills

- A misjudgment about the benefits and particularities of a BIM delivery process

- Ambiguity about the BIM Manager's position within an organization

> **HOW TO OVERCOME THIS ISSUE:**
>
> Understand the reasons behind the lack of engagement by upper management.
>
> Learn how upper management makes decisions—What is their source of information? Who do they trust and why?
>
> Engage the key decision makers in a discourse about BIM—avoid being technical. Put yourself in their shoes.
>
> Ensure BIM becomes an integral part of business conversations; run regular information sessions with upper management.

Lacking Support from the Top

Management's lack of understanding of BIM results in a bigger problem: the lack of adequate support they provide!

BIM Managers are often left with insufficient authority to facilitate change and to set up BIM in a sustained manner. Still, it is the BIM Managers who get blamed for the lack of progress if change does not happen or if it does not occur quickly enough. This problem highlights a crucial quality BIM Managers need to possess when they guide an organization's transition to BIM. They need to be able to make and communicate a compelling business case for BIM to their leaders.

Making a business case is highly relevant in order to receive adequate resources and time allocations for the management process of BIM. Resources in this sense refer to (training and coaching of) staff, but also to a dedicated budget for BIM-related purchases (software/hardware, BIM library objects, etc.). Time allocations refer to both the BIM Manager's time away from pure project work (e.g., in order to focus on BIM standards development) as well as an understanding that staff in general need to be given time for ongoing training, mentoring, or other forms of engagement with BIM. Gustav Fagerström is a Design Technology Specialist who gathered his experience working across several different fields and prominent firms such as Buro Happold, UN Studio, and KPF. He describes this issue as follows: "Oftentimes there is time to either do the work your clients are directly paying you for, or to develop non-project-specific best practices and tools, but not both."

> **HOW TO OVERCOME THIS ISSUE:**
>
> Learn to gain trust by upper management.
>
> Prove the value you add as a specialist within your organization.
>
> Demonstrate that BIM is not an overhead by a platform for innovation.

Be clear about your BIM-related responsibilities and accountabilities and those of others.

Highlight what it is you are doing and what support you require to do your job well.

Make (a) business case(s) for the change you wish to enable.

Set up and articulate a BIM and Design Technology–related annual budget.

Becoming a Manager

Issues related to a lack of support from the top are a double-edged sword. While BIM Managers agree that the lack of support from upper management is a problem, this problem could be seen in a different light: Any professional who puts "Manager" in his or her title should better be able to deliver on that promise.[8] The point made here is that BIM Managers can—to a degree—be blamed for their peculiar situation: Their technical (or BIM software) knowledge is not always matched by equivalent leadership skills required to act in a management role. BIM Managers need to know how to manage process and change, and not just software. The BIM Manager Certification courses mentioned in Chapter 1 are a pathway into more structured skill development for BIM Managers, but there is more.

Upper management laments the lack of communication skills of BIM Managers who struggle to articulate their specific needs. Independent of the level of authority provided to BIM Managers within an organization, they are still accountable to enable upper management to make informed decisions about BIM.

HOW TO OVERCOME THIS ISSUE:

Improve your communication skills—You are more likely getting support if you can clearly articulate what you need.

Attend Management seminars that focus on business aspects of your work.

Aim for gaining an "industry-accepted" BIM Manager Certification.

Establish benchmarks and referring timelines or budgets against which they can be measured.

Apply management techniques to your own workflow; introduce BIM key performance indicators (KPIs).

Work together with upper management on business plans and budgets that relate to your work.

Learning to Lobby

Years of experience in a 700+ staff design firm have taught Toby Maple, global Design Technologies Leader at HASSELL, a thing or two when it comes to facilitating change. According to Maple, in order to win over the skeptics, a BIM topic is best not tabled impromptu at an executive meeting; it is better to strategically increase management's understanding and seek common alignment from the relevant players beforehand. Toby explains:

"Management's understanding of BIM and BIM processes are probably not to the same technical level as the BIM Manager. One key tactic for getting multiple parties aligned, understanding and supporting you is to have nontechnical discussions with them one-on-one—before you go into the room where you seek an outcome. If you can arm them with a good understanding of what the issues are beforehand, you certainly increase your chances of reaching a consensus. People don't want to look ill-informed and not able to make a decision in a meeting!"

"In terms of the actual process of providing decision support: You need to be able to articulate to management the value of improving on the current situation, as well as to highlight the negatives to adoption; you need to define the business imperatives to make certain decisions, because of cost, resourcing, or maintenance of different applications. Once people realize that your request is not simply a passionate or emotional plea, that the issues you raise affect the business financially, that realization can result in a Eureka moment for management. These people often become your greatest advocates."

Figure 2–4 Federated Revit Models in Navisworks for coordination reviews by HASSELL.
© HASSELL

An organization's leadership is well advised to adopt a strategic approach to implementing BIM that tackles Change Management on a business and cultural level. The BIM Manager is the key advocate to allow this change to unfold, supported by upper management and in strong reference to the goals and objectives by key staff such as project leaders and practice innovators.

The Inside Man

Following Toby Maple's suggestions for engaging management strategically, another instrument to break the ice between BIM management and an organization's leadership is to have an individual from within the leadership group acting as the BIM liaison—a senior staff member who not only enjoys the trust of upper management, but who also has a technical edge. He or she can help facilitate dialogue and engagement between the more technically inclined BIM proponents and the directorial/principal level. The advantage for BIM managers is to have a strong partner to go to who is in regular contact with decision makers in a firm. The disadvantage is the BIM Manager's dependency on that liaison which, to a degree, may stymie the BIM Manager from gaining direct access and respect from the leaders. The reliance on a liaison may also turn out to be problematic if that person is drawn away by other commitments and BIM- or technology-related issues remain unresolved over an extended period of time.

Ultimately, it should be the BIM Manager's goal to have the ear of upper management directly and to aspire to becoming part of that management group. Current examples in practice hint at the increased significance organizations place on the BIM Manager's role. A career pathway toward Design Technology Leader, or even Practice Director is a plausible development.

Selling Value Back to the Business

Strategic BIM Manager and Associate Director at HDR Rice Daubney, Stephan Langella is one of a lucky few. When discussions about people's roles came up within his organization, Langella's boss delivered a clear message to his peers: The BIM team is not an overhead; instead, the employment of someone like Stephan Langella is a strategic investment into the future of the business. Langella accounts:

"You have to stop the culture where you accept the business referring to you as an overhead. You need to claim your position as a strategic investment. The problem is that some within a business believe that utilization equals profit; and that is not true. In order to demonstrate your value you need to show tangible outputs that others can engage with. You need to sell your value back to the business! The smart BIM Manager needs to know how to do that. Smart BIM Managers need to articulate business cases; they need to have a business focus. Otherwise you may get pushed into a corner. If all you have left to do is 3D documentation and fire-fighting, you have lost control! We keep implementing the following strategy on our organization: Every single project has a person assigned to it who is responsible for BIM. That person doesn't necessarily have to be a capable BIM person, that person has to be a capable manager."

As Langella's example highlights, BIM Managers often struggle to provide proof of their value to their organization. How far does the investment into BIM pay off? What is its business case (from project to project)? How can

Figure 2–5 Best Use of BIM for Design, Drama, and Excitement at Build Sydney Live 2013, BIM Coordination Workshop result by HDR Rice Daubney.

© HDR | Rice Daubney and Obayashi Corporation

one measure successes or challenges of BIM and validate performance on projects or across an entire organization? How can the Return on Investment of BIM be maximized?

By nature, BIM proponents argue for an increase in output quality of documentation, construction planning, the construction process both onsite and offsite, and higher-quality data at handover and commissioning. The effort applied to achieving higher-quality output differs from stakeholder to stakeholder as roles and responsibilities can shift between traditional and BIM-style approaches. Finally, fees are mostly still determined by clients and project managers based on outdated traditional methods of delivery. They seldom reflect the nature of upfront planning and coordination associated with BIM and they rarely take into account the potential of BIM to facilitate downstream savings for clients. All of the above arguments need to be taken into consideration by BIM Managers when defining their business cases and the associated Return on Investment (ROI).

While there exist simplified formulas[9] or even "online tools"[10] to measure BIM ROI, the prerequisites for achieving efficiency gains from BIM are often more multifaceted. These benefits depend on a large number of factors and they differ throughout the various trades, markets, and maturity levels of BIM uptake. They also depend on third-party requirements such as BIM mandates or specific BIM-related policies. Implementation costs for software, hardware, training, or content (creation) can represent a significant hurdle that prevents some from going for BIM in the first place. Those who do are still left with a steep learning curve and ongoing costs for implementation (and in particular for managing change). Design Technology expert Gustav Fagerström shares his personal views on the pros and cons of the project-related utilization options:

"It is a key conundrum for the management of Design Technology within any AEC organization whether it is preferable to remain an overhead cost entity (more R&D headspace, less project integration, thereby potentially less easily demonstrable relevance to the practice bottom line and ROI), or an integrated project team entity (little to no headspace for R&D beyond specific project applications, integrated fully into the billable work of the organization and essential to delivering the product.) Either model has its own challenges in terms of quantifying and validating the associated upfront cost as well as the ROI."

HOW TO OVERCOME THIS ISSUE:

Benchmark BIM-related expenses and activities and establish a clear set of KPIs against which they can be measured.

Report these KPIs regularly to upper management and discuss their implications.

Work with project team leaders on their delivery strategy; weave in your BIM knowledge to increase efficiencies.

Explain to upper management what you manage, and what you fail to achieve with BIM; provide the reasons behind both.

Establish an annual Design Technology and BIM budget from which to work; get it funded by upper management.

Involve upper management in decision-making processes related to the financial aspect of BIM.

Shift upper management's mindset away from seeing BIM as an overhead and toward an investment into the future of your organization.

Overcoming Change Resistance and Managing Expectations

Director of Digital Practice, Josh Emig of Perkins+Will states, "While integrated delivery is growing, the majority of projects in our industry still use delivery methods that reinforce disciplinary silos, narrow self-interest, risk aversion, information hoarding, and a 2D mindset."

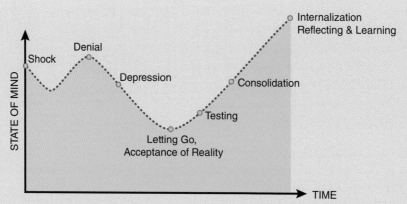

Figure 2–6 After J. Hayes, *The Theory and Practice of Change Management*, 2010.[12]
© **Dominik Holzer/AEC Connect**

So far we have looked at change management on an organizational level and how it is influenced by organizational and leadership culture. This section addresses change resistance as it occurs in groups and with individuals. The key impact on our behavior occurs on an emotional and attitudinal level.[11] Fear plays a role that feeds into resistance to change—fear of the unknown, fear related to financial loss, or fear of becoming redundant. Organizational Behavior Research tackles this issue by identifying several states of mind an individual undergoes when confronted with transition.

Winning over leadership when it comes to BIM is without doubt a crucial precondition to effect change within an organization. Yet, when it comes to addressing change resistance on an individual level, there is another group that requires attention: Middle Management, and in particular project team leaders.

That BIM Thing Looks Amazing, Just Not on My Project!

According to BIM Manager Jan Leenknegt at Bjarke Ingels Group (BIG), "Everyone comes to a Copernican shift in documentation technology with their own approach, their own assets and fears. You get the whole range of responses—in times of challenge the real personality shines through."

BIG's Jan Leenknegt accentuates a primary concern of BIM Managers when facilitating change: Winning over the design teams and those running the projects with those teams. BIM Managers benefit greatly from the support they get from the top. At the same time it is actual BIM output and the end user comfort on a team level that reflects the success or the shortcomings of any BIM implementation effort most directly. Jan explores this further:

"In general, the best way to effect change (with the partners, design team members, and especially with project leaders) is to work through actual results rather than through "BIMspeak" or boilerplate BIM slide shows.

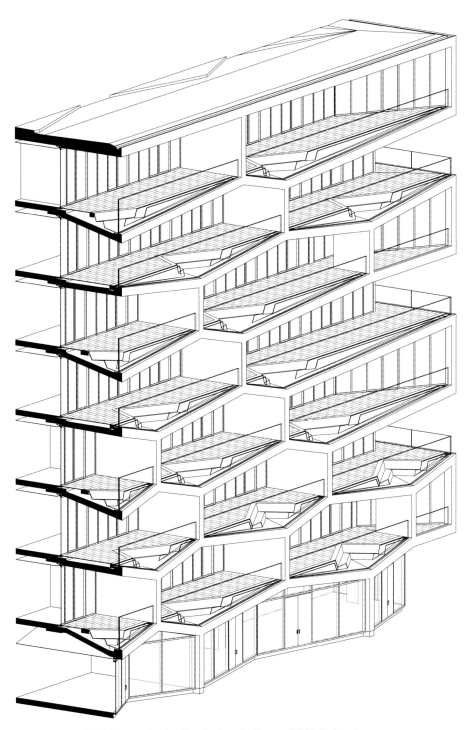

Figure 2–7 BHS Project by the Bjarke Ingels Group (BIG): Balconies.
© BIG

Figure 2–8 BHS Project by the Bjarke Ingels Group (BIG): Close-up Render.
© **BIG**

A sharp-looking finished drawing set or a well-performing interoperability workflow will win over more BIM skeptics than all the MacLeamy curves in the world.

Another way, [to gain the support of] the design teams, is to always prioritize end user comfort over BIM management comfort. Example: A Revit template with a long list of Worksets may look professional and ready for all possible scenarios from a BIM management point of view, but it will, in most cases, only add extra assigning work for the design teams. We start our projects with one Workset for all modeled elements ("Building"), and add more only when necessary. Most projects do perfectly fine with that one Workset through halfway of the Design Development stage."

People resist change. They do so in a context where they see their well-established "formula for success" being jeopardized by approaches they aren't familiar with and that they haven't tested yet. Change resistance is a

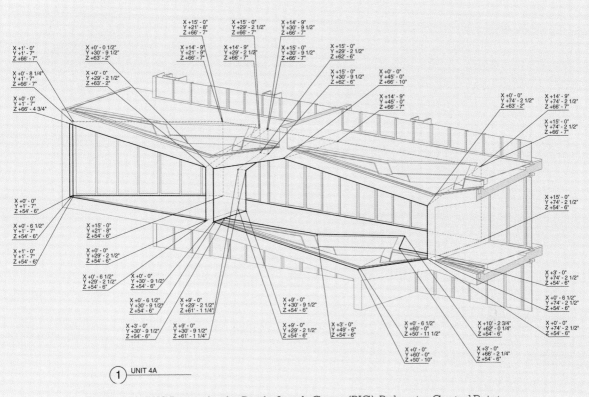

X +1' - 0"
Y +1' - 7"
Z +66' - 7"

X +0' - 8 1/4"
Y +1' - 7"
Z +66' - 7"

X +0' - 0"
Y +1' - 7"
Z +66' - 4 3/4"

X +0' - 0 1/2"
Y +30' - 9 1/2"
Z +63' - 2"

X +0' - 0"
Y +29' - 2 1/2"
Z +63' - 2"

X +15' - 0"
Y +21' - 8"
Z +66' - 7"

X +15' - 0"
Y +29' - 2 1/2"
Z +66' - 7"

X +14' - 9"
Y +30' - 9 1/2"
Z +66' - 7"

X +14' - 9"
Y +21' - 9"
Z +66' - 7"

X +14' - 9"
Y +29' - 2 1/2"
Z +66' - 7"

X +15' - 0"
Y +30' - 9 1/2"
Z +66' - 7"

X +15' - 0"
Y +29' - 2 1/2"
Z +62' - 6"

X +15' - 0"
Y +30' - 9 1/2"
Z +62' - 6"

X +0' - 0"
Y +30' - 9 1/2"
Z +62' - 6"

X +0' - 0"
Y +45' - 0"
Z +66' - 10"

X +14' - 9"
Y +45' - 0"
Z +66' - 7"

X +0' - 0"
Y +74' - 2 1/2"
Z +63' - 2"

X +14' - 9"
Y +74' - 2 1/2"
Z +66' - 7"

X +15' - 0"
Y +74' - 2 1/2"
Z +66' - 7"

X +0' - 0"
Y +1' - 7"
Z +54' - 6"

X +0' - 6 1/2"
Y +1' - 7"
Z +54' - 6"

X +1' - 0"
Y +1' - 7"
Z +54' - 6"

X +15' - 0"
Y +21' - 8"
Z +54' - 6"

X +0' - 0"
Y +29' - 2 1/2"
Z +54' - 6"

X +0' - 6 1/2"
Y +29' - 2 1/2"
Z +54' - 6"

X +0' - 0"
Y +30' - 9 1/2"
Z +54' - 6"

X +0' - 6 1/2"
Y +30' - 9 1/2"
Z +54' - 6"

X +0' - 0"
Y +29' - 2 1/2"
Z +61' - 1 1/4"

X +9' - 0"
Y +29' - 2 1/2"
Z +61' - 1 1/4"

X +3' - 0"
Y +30' - 9 1/2"
Z +54' - 6"

X +9' - 0"
Y +30' - 9 1/2"
Z +61' - 1 1/4"

X +9' - 0"
Y +29' - 2 1/2"
Z +54' - 6"

X +9' - 0"
Y +30' - 9 1/2"
Z +54' - 6"

X +3' - 0"
Y +49' - 6"
Z +54' - 6"

X +0' - 0"
Y +60' - 0"
Z +50' - 11 1/2"

X +0' - 0"
Y +60' - 0"
Z +50' - 10"

X +0' - 6 1/2"
Y +60' - 0"
Z +50' - 11 1/2"

X +10' - 2 3/4"
Y +62' - 0 1/4"
Z +54' - 6"

X +0' - 0"
Y +66' - 2 1/4"
Z +54' - 6"

X +3' - 0"
Y +66' - 2 1/4"
Z +54' - 6"

X +15' - 0"
Y +74' - 2 1/2"
Z +54' - 6"

X +3' - 0"
Y +74' - 2 1/2"
Z +54' - 6"

X +0' - 6 1/2"
Y +74' - 2 1/2"
Z +54' - 6"

X +0' - 0"
Y +74' - 2 1/2"
Z +54' - 6"

1 UNIT 4A

Figure 2–9 BHS Project by the Bjarke Ingels Group (BIG): Balconies Control Points.
© BIG

natural instinct to protect oneself from risk of failure. Knowledge acquisition in the construction industry is often based on empirical observation and exposure to processes as they unfold on precedence projects. The advancement of knowledge vocationally stands in stark contrast to the system-oriented approach taken in other industries, such as car manufacturing or aerospace.[13] The increased use of BIM as the predominant method of delivering projects jeopardizes previously existing success formulas applied by middle management who usually occupy team-leader roles.

HASSELL's Toby Maple shares an insight related to this topic:

"On a project level, if we consider a BIM skeptic principal or team leader, it is crucial to get good people around them, who have delivered projects before successfully using BIM; they value the opinion of their peers. As Design Technologies Leader they don't necessarily value my opinion from a project delivery standpoint, even though I have been implementing BIM for over ten years. They will trust their fellow staff whom they've worked with before. They will be more comfortable asking them questions such as: How can I achieve 3D coordination? How can I integrate my markup's into the model? HASSELL is luckily now in a position where we have a network

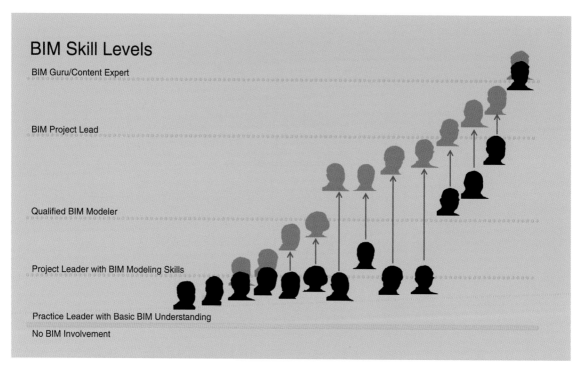

BIM Skill Levels

BIM Guru/Content Expert

BIM Project Lead

Qualified BIM Modeler

Project Leader with BIM Modeling Skills

Practice Leader with Basic BIM Understanding

No BIM Involvement

Figure 2–10 BIM skill level development on an individual level.
© **Dominik Holzer/AEC Connect**

of BIM support personnel for staff. We have been fortunate to have delivered profitable BIM projects very early on which gave us enough success stories for BIM to become infectious—from then on people started engaging with BIM processes on their projects and asking: *How can I use BIM to do my job more effectively?*"

Support for BIM not only takes into consideration individuals' personalities, but it should also take into account the desired level of change for each individual. As an example, when looking at existing and desired skills across an organization, the BIM Manager should map out different levels of BIM-related knowledge to then determine each staff member's desired pathway for change.

The mentality of those who do not adapt to the changing circumstances in the workplace needs to shift in order for them to get rid of old habits. BIM Managers need to take on a coaching role; easing the burden of transition for those who struggle and empower them to become part of the BIM implementation process. Well-formulated people skills are a prerequisite for BIM Managers who provide support to others whose focus lies less on technology than other aspects of their work. In the context of cultural change, the ability to go beyond the kind of silo thinking prevailing across the construction industry is a pivotal asset for BIM Managers. Where others often only see their part of the equation, BIM Managers need to provide vision and guidance that cuts across professional boundaries, thereby embracing the whole-of-life dimension of BIM.

HOW TO OVERCOME THIS ISSUE:

Overcoming change resistance requires acknowledging that it exists in the first place.

Proper BIM management requires an in-depth understanding of change management.

Communicate the cultural relevance of BIM to your organization on a number of levels:

- A solid and ongoing dialogue about technology and BIM, and business goals with upper management
- Regular information sessions with project leaders, understanding their point of view and their deliverables
- Regular and frequent mentoring/monitoring with those who author or coordinate information based on BIM

Facilitate a transparent discourse about the advantages and challenges associated with project delivery using BIM.

Realistically judge each individual staff member's pathway for applying BIM and complement it with related training and mentoring.

Change management has a vast array of issues associated with it; Chapter 2 of this series is dedicated entirely to this topic.

Bridging the "Us vs. Them" Schism

"…the extreme positions of technophobia and technophilia provide a useful dichotomy that animates and propels design thinking."[14]

We have all come across labels such as "the BIM guys" or the "Tech People." There exists a cultural schism associated to the implementation of BIM that is deeply rooted in the history of technological advance in practice.

The use of high-end technology in professional practice has progressed since the advent of personal computing in the 1980s and from the mid-1990s onward it was greatly amplified by the added connectivity facilitated through the World Wide Web. Nowadays, the application of high-end technology has become pervasive in the developed world both in our private lives as well as for our work; the boundaries between specialist and common operator are progressively blurring.

Within organizations the progressive introduction of technology has, at times, led to an "us vs. them" mentality. There are those who see the use of technology less as an opportunity than as a threat to their professional integrity; a distraction to the core of their professional activities. As a result, those who engage with technology can easily get pigeonholed as "tech people" (or "techies") with an associated career path that keeps them separate from others. This stigmatization may at times be the typical context for a BIM Manager's career. Such a stigma may be welcomed by some who want to specialize and who see this as a distinction on their professional pathway. Others may struggle accepting that BIM helps their careers.

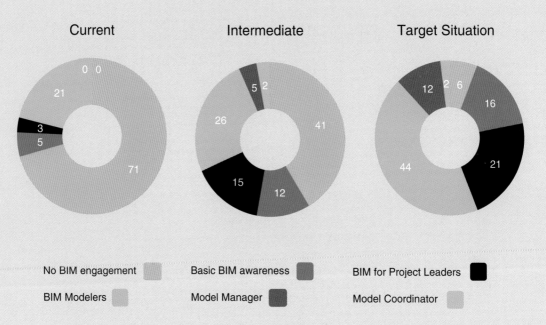

Current   Intermediate   Target Situation

No BIM engagement   Basic BIM awareness   BIM for Project Leaders

BIM Modelers   Model Manager   Model Coordinator

Figure 2–11 Strategic BIM skill roadmap considering individual strengths and organizational goals.
© Dominik Holzer/AEC Connect

As an example, there are designers who do not wish to develop their BIM skills as they fear the label of "Model Manager" or "BIM Architect." Engineers may be critical about adopting BIM as it pulls them stronger into design documentation and delivery—a role that for many instead sits with specialist drafters and detailers. Contractors may prefer to do things in the way they have done them in the past. The use of BIM suggests disrupting their established principles, which in return is perceived by contractors as a substantial risk. Those propagating BIM management within contracting firms therefore continuously fight to demonstrate that the added benefits of BIM outweigh the risks.

The type of tensions mentioned above can lead to organizational friction and inertia. Colleagues may—at first—be interested in technology, change, and the promise of more efficient work methods. When it comes to aligning their known ways of working with BIM processes they may struggle to accommodate such change. Techno-skepticism and Technophobia among certain staff may be the result and lines get drawn between technology protagonists and those who wish to have little to do with it.

Developing a Network

"Change in a large, diversified, geographically distributed firm requires strong networks for both initial and long-term success," says Josh Emig, of Perkins+Will.

No BIM Manager works in isolation. As in any other organizational setting, BIM Managers constantly interact with their peers to advance their desired goals. BIM Managers engage regularly with the executive level and they interact with project teams on a day-to-day basis. Depending on the size of any organization and its geographical distribution, BIM Managers should keep an eye on building up partnerships and networks across their organization that help them to effectively expand their sphere of influence. One of the worst occurrences in larger organizations is the segregation into local clusters of interest where synergies get lost and uncoordinated approaches to BIM flourish. Josh Emig, Director of Digital Practice at the design firm Perkins+Will, shares his experience:

"Perkins+Will is a design firm of 1,600 people, working in ten or so market sectors, across 25 offices, globally. Individual office size ranges from 30 people to over 200 people. The growth of Perkins+Will over the last 15 years has been an equal split of organic growth and acquisitions. Each acquisition joins the firm with its own culture, personalities, and ways of working. Lastly, Perkins+Will has a distributed corporate structure, which is to say that there is no central Perkins+Will headquarters. Our corporate leadership is distributed across five cities in North America, while our board membership represents an even broader set of localities globally.

Several years ago, when we went about rethinking our technology efforts, our design technology organizational structure consisted of two groups: a corporate group, focusing primarily on the implementation and support of design software, and local project-based 'design technology leaders.' This structure was actually in place for good reason: It allowed for central coordination, as well as project-based champions, for initial BIM implementation and support that began in 2006. Over time, however, as the firm grew and BIM adoption became more mature, this structure began to show weakness."

"While the firm did not suffer for talent, the corporate group had become disconnected from projects, project leadership, and office leadership, and project technology leaders who, while effective on projects, were similarly disconnected from broader efforts at the firm-wide scale. What was needed were more and stronger connections between office networks and the firm-wide network.

Figure 2–12 High performance buildings, like Perkins+Will's Atlanta office at 1315 Peachtree St. in Atlanta, require diverse skill sets and technology perspectives to execute successfully. Strong, diverse internal social and organizational networks are a key component of building successful teams.

© Raftermen Photography

Figure 2–13 Perkins+Will major, multi-office project BIM planning incorporates perspectives from various domains in network: overall project manager, firm-wide BIM leader, office BIM managers, and project BIM managers representing several Perkins+Will and consultant offices.

© Perkins+Will/Josh Emig

Through simple organizational shifts, we created technology leadership positions in each office—effectively a local technology leader who reports to office leadership, coordinates the efforts of the project-based technologists, and forms a network tie to our smaller, corporate design technology group, now called 'Digital Practice.' These office-based technology leaders are also key communicators of technology capability, strategy, and risk management to our project leadership, which is perhaps the most critical constituency in any change program. Lastly, they are often the best 'ears' in the firm because they are privy to project details that are invisible to the corporate team because they are 'flying too high' and invisible to project teams because they are 'in the weeds.'

In each of these domains, people are engaged with efforts of the others—project-based technologists often work on firm-wide knowledge and resource development efforts. Corporate Digital Practice team members engage on project work. Office design technology leaders work in the three domains simultaneously. Most importantly, the communications of these three domains overlap continuously through various standing calls, through project efforts (both internal resource development and external, client-facing projects), as well as through friendship and 'peer support'."

Tips and Tricks

This concluding section of Chapter 2 offers some in-depth support for running BIM audits and for setting up in-house BIM Workshops. These instruments provide BIM Managers with a better understanding of their organizational context and they help to raise the BIM knowledge level of key decision makers.

The Design Technology and BIM Audit

There are a number of ways for BIM Managers to address cultural and skill issues related to Change Management. An audit is a great starting point to bridge the gap between an organization's aspirations related to technology use, and more specifically BIM, and successful implementation. The audit can either occur in-house, or it can be extended to include key collaborators of that organization. The purpose behind the audit is twofold.

First, it will offer the BIM Manager a stock take of the organization's BIM capacity (summary of skills) as well as revealing the aspirations of staff from the leadership level all the way to the operators on the floor. The BIM Manager can then identify gaps between the current BIM knowledge level and the desired BIM maturity within the organization. Identifying those gaps will assist the BIM Manager in developing a roadmap for future implementation.

Second, the BIM audit is about empowerment. The fact that the audit takes place is in itself a message to staff that their views are valued. In addition, communicating findings from the audit to those involved (and even those not yet involved) will engage them with the topic. The insights provided by the audit will allow the BIM Manager to engage leadership with confidence about staff sentiment and the current skill level among them. It cuts out a good proportion of "second guessing" by the BIM Manager and it helps to fine-tune the development of future pathways for implementation.

One cannot assume that staff will have the answers when it comes to the right implementation strategy. Ultimately, the BIM Manager is the key person tasked with developing a vision and direction for the change that is required. Still, responses from those affected most by BIM in their day-to-day work are invaluable when it comes to Change Management and BIM implementation.

Set Up and Run a Design Technology/BIM Audit

The audit itself should consist of three major parts.

The first part is about gathering people's individual feedback during semi-structured, face-to-face interviews with selected staff. Here the BIM Manager asks employees about their experiences and their expectations of using BIM. Transcripts or even recordings are useful aids to help the interviewer remember what gets said. In case the interviewing BIM Manager uses any method to record what is said, he or she must ensure that no office policies or other work-right-related clauses are violated. Also, anonymity of the interviewees needs to be

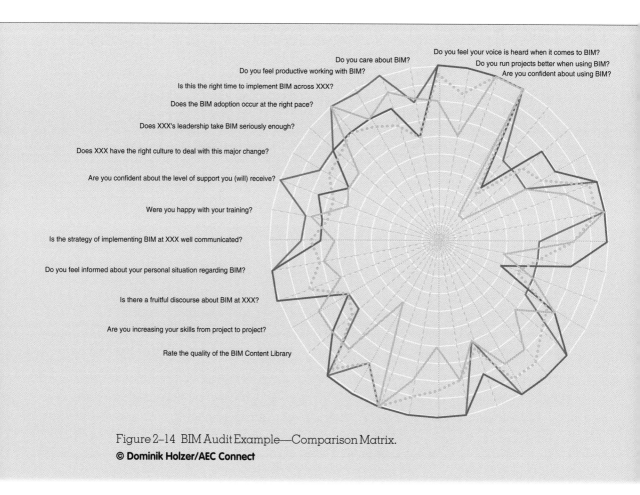

Figure 2–14 BIM Audit Example—Comparison Matrix.
© **Dominik Holzer/AEC Connect**

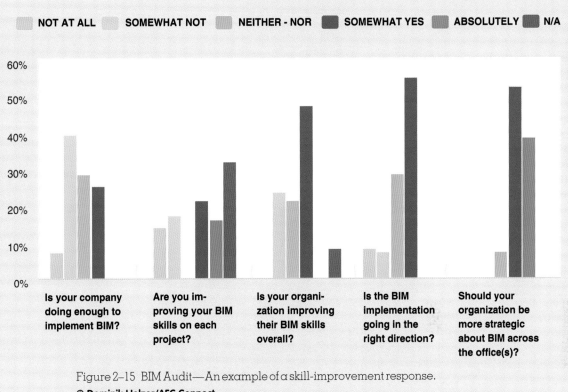

NOT AT ALL   SOMEWHAT NOT   NEITHER - NOR   SOMEWHAT YES   ABSOLUTELY   N/A

Is your company doing enough to implement BIM?

Are you improving your BIM skills on each project?

Is your organization improving their BIM skills overall?

Is the BIM implementation going in the right direction?

Should your organization be more strategic about BIM across the office(s)?

Figure 2–15 BIM Audit—An example of a skill-improvement response.
© Dominik Holzer/AEC Connect

guaranteed in order to establish trust between the interviewer and the interviewee. In some cases, an organization may prefer to invite an external party to conduct the interviews.

Complementing the interviewing process, the BIM Manager can find out more via a questionnaire that asks respondents to quantify certain aspects related to BIM and technology. This second part of the audit will allow the BIM Manager to graphically represent responses on charts. Such charts are highly useful in order to communicate findings to an organization's leaders and others.

The third major part of a BIM audit is to work in groups to run collaborative brainstorming sessions where employees are able to voice their views and develop out-of-the-box ideas related to BIM. These sessions are usually highly interactive and they may already include a component where staff gets exposed to some of the feedback gathered during the audit. Raising consciousness, increasing understanding, and making individual staff aware that their concerns and ideas are noted by others lead to social interaction and empowerment.[15] The group dynamic emerging from any audit should be expanded and developed into regular BIM user sessions.

What Should Be Asked during the Audit?

A BIM audit should obviously help determine the current skill level of staff when it comes to delivering projects using BIM. In addition, the audit should engage staff on a personal level in order to determine the mood on the floor and broader aspirations. The BIM Manager should organize the questions in a manner that includes more general questions upfront, to then become more specific toward the end. Ideally, questions should cater to those already confident in the use of BIM as well as to those not engaged yet. Further, questions should query the level of confidence interviewees have in the BIM approach taken by their organization. How well are they informed about what is going on? Do they feel involved in the uptake/implementation of BIM? It proves useful to ask participants during the audit about their expectations and if those are matched by what they have experienced in terms of BIM exposure.

SOME QUESTIONS MAY INCLUDE:

How can/could it support your daily (design) practice?

What are the obstacles to using BIM in an efficient way? Is there anything holding you back?

Why should you engage with BIM at your organization? Why wouldn't you?

What would you put in your miracle toolbox; what would make your work so much easier?

Describe the way you exchange BIM data with your consultants from design to construction.

How easily do you adapt to unfamiliar technology?

What training would you require most for working in a BIM environment?

To what extent does BIM affect the documents and data output capability within your organization?

Has progress been made with its implementation?

What would you like to achieve through BIM?

How often do you get asked by external design parties to provide information in BIM?

As much as in-house audits offer a rich view into an organization's technology uptake, BIM Managers should also consider including external organizations in such activities. It can be invaluable to the positioning of an organization's BIM strategy to have a good understanding about the BIM experience of their most immediate collaborators.

Due to obvious business constraints, external BIM audits usually cannot be conducted with the same level of detail as internal ones. External audits can nevertheless reveal important aspects to be considered as part of an organization's internal Change Management process. They paint a more detailed picture of the market in which an organization is operating. Such knowledge can help with the alignment between internal work processes and external factors that are crucial for the successful delivery of projects using BIM across multiple organizations.

Table 2.1 BIM Audit—Typical example of an audit questionnaire.

QUESTIONS

General

	absolutely	somewhat yes	neither - nor	somewhat not	not at all	NA
Do you care about BIM?						
Do you feel productive working in BIM?						
Is BIM "the right way" for delivering projects?						
Is this the right time to implement BIM across XYZ?						
Does the adopting happen at the right pace?						

Level of Confidence

	absolutely	somewhat yes	neither - nor	somewhat not	not at all	NA
Are you confident XYZ's leadership takes BIM seriously enough?						
Does XYZ have the right culture to deal with this major change?						
Are you confident with your ability to migrate from CAD to BIM software?						
Are you confident about the level of support you receive?						

Being Informed

	absolutely	somewhat yes	neither - nor	somewhat not	not at all	NA
Were you happy with your BIM training?						
Do you feel informed about the steps XYZ is taking (to implement BIM)?						
Do you feel well informed about XYZ's internal BIM support?						
Is the strategy of implementing BIM at XYZ well communicated?						

Feeling Involved

	absolutely	somewhat yes	neither - nor	somewhat not	not at all	NA
Do you feel involved in the BIM process on your project?						
Do you feel part of XYZ's BIM progress?						
Do you feel your voice is heard when it comes to BIM?						
Is there a fruitful discourse about BIM at XYZ?						

Matching Expectations

	absolutely	somewhat yes	neither - nor	somewhat not	not at all	NA
Does BIM at XYZ meet your expectations?						
Are you happy with your own learning curve?						
Do you think starting BIM was worthwhile?						
Do you enjoy working with BIM?						

Output Quality

	absolutely	somewhat yes	neither - nor	somewhat not	not at all	NA
Has BIM had an impact on the quality of your documentation output?						
Do you run projects better when using BIM?						
Does using BIM make your life easier?						
Does BIM allow you to increase the quality of your design?						

Progress

	absolutely	somewhat yes	neither - nor	somewhat not	not at all	NA
Is XYZ doing enough to implement BIM?						
Are you improving your BIM skills on each project?						
Is XYZ improving their BIM skills overall?						
Is XYZ's BIM implementation going in the right direction?						

© Dominik Holzer/AEC Connect

For the above reasons, it is useful to find out how far advanced collaborators are with the setup of their in-house strategy. How well established are their workflow protocols? Do they implement well-formulated standards? Do they work toward accepted industry frameworks and guidelines? Would they like to approach certain aspects of collaboration differently, given the opportunities BIM provides? How do they approach document and data management in the context of BIM? What value add do they perceive within the context of BIM? How would they usually split BIM element authorship across various stages of design, engineering, documentation, and delivery?

A greater awareness of these factors allows BIM Managers to reduce second guessing when it comes to the establishment of collaboration workflows. It further allows the BIM Manager to communicate preferences to decision makers within an organization when it comes to the selection process for collaboration during tender.

The BIM audit does not merely assist the BIM Manager in understanding the cultural Change Management within an organization. The audit can be seen as a stepping stone toward the development and/or adjustment of BIM standards, BIM content creation framework, and ultimately also the setup of cross-disciplinary BIM Execution Plans.

Change Management Workshops and Seminars

With the audit as a first step, more support is required to assist an organization with Change Management. Learning the use of new software itself is seen by some as the key move toward adopting BIM. This classic misconception originates in the belief that BIM is some sort of 3D CAD and all that is needed is therefore an understanding of generating, manipulating, and checking/coordinating 3D models. As seen in Chapter 1 of this publication, Best Practice BIM draws a very different picture.

Adopting BIM successfully requires a sound understanding of information management across the project supply chain. Next to a different approach to project setup, BIM necessitates changes to established workflows within an organization and beyond. Further, BIM offers a range of opportunities for data association and integration with processes that can be entirely new to an organization (and the expanded project team). There is no "typical" BIM workflow; it strongly depends on what information requirements are predominant on a project-by-project basis. Such an approach may be alien to organizations who traditionally were used to delivering more ad-hoc outcomes under a lot of time pressure and with only limited consideration about data integration and management.

Bearing in mind the opportunities mentioned above, organizations usually still depend on 2D document delivery to fulfill their contractual obligation. Typical plan/section/elevation output that traditionally stemmed from the drafting process in CAD increasingly gets generated in semiautomated processes using BIM. Manipulation of 2D output is often no longer directly possible for project architects, project engineers, or project/design managers. Accessing design in order to update 2D document output does not occur as directly in BIM as it used to in CAD. The interconnectedness of information inherent to BIM makes it difficult to accommodate local or ad-hoc changes to document output. The entire setup of graphic standards for the visual feel of document output is best determined upfront and not at the end of the process toward finalizing documents for submission.

For those who are not (yet) familiar with this typical BIM workflow, access to crucial documentation output is suddenly kept at arm's length. As a result, there exists the danger of disempowerment and even frustration among those who traditionally used to run projects with hands-on control about the output. If they apply a traditional (CAD) mindset to running projects, BIM will not yield the promised increase in efficiencies. On the contrary, the use of BIM is likely to delay progress among the project team, or—at times—the team may even revert back to using CAD in order to meet urgent project deadlines. In engineering practice, BIM is changing the role distribution between the drafters and the engineers. Depending on local (and market) circumstances, it is not uncommon for there to exist a clear split between the duties of a draftsperson (for document creation) and those of an engineer (for design simulation, analysis, and validation). When using BIM, there is a chance that engineers are increasingly becoming involved in the documentation process, as it now more easily ties into various analysis processes. As a result, some see the role of the draftsperson in danger when considering the future development of BIM among engineers.

The BIM Manager needs to address these concerns and help educate key project staff about the intricacies associated with various BIM workflows. Managing expectations among staff is a fundamental aspect of Change Management. The BIM Manager thereby helps those new to the BIM workflow to understand the major differences to traditional approaches of project delivery. The Change Management seminars should neither focus much on the act of documenting a project in BIM nor to the software used to do so. They should cover other related topics such as the project context, project initiation, project progress, and the handover of information to downstream parties.

PROJECT CONTEXT	BIM Manager Tasks
Interpreting BIM clauses in project briefs	Make project leaders aware of the typical BIM deliverables in project briefs to ensure they avoid signing off on services that may sit outside your organization's typical scope.
Understanding the impact of contract procurement	Highlight how various contract procurement models impact on the team's ability to share data via BIM.
Understanding legal obligations associated with BIM delivery	Advise the project leader on how to consider these topics as part of the project setup.
The relevance of national guidelines or key BIM requirements of select clients	Inform staff about the most relevant national BIM requirements (as per mandates of public sector government bodies). Project teams ought to understand the key implications such mandates or requirements may have on their deliverables.
Individual staff's BIM skill levels and development thereof	Get an overview about the range of BIM skill levels on the floor. Envisage the desired skill levels across the organization and identify how to educate individual staff toward their desired knowledge level.
The relevance of in-house BIM Standards	Highlight the need for such standards and explain to staff how they affect their day-to-day BIM delivery processes.

PROJECT STARTUP	BIM Manager Tasks
Project startup and work throughout the consecutive stages in BIM	Assist project teams in determining the most appropriate moment for them to start using BIM on a project. Help to map out a roadmap for the use of BIM throughout the remaining project stages.
Explain the concept behind digital tool ecologies and interoperability	Be aware of the most appropriate path to connect different tools for design, documentation, visualization, analysis, and so forth. Assist project teams in defining related tool ecologies, file transfer, and data-handover processes.

PROJECT STARTUP	BIM Manager Tasks
Discuss project team selection in relation to BIM	Consider the skill sets required to use BIM successfully, and assist project leaders in selecting staff with complementary skill sets in order to get the right mix for their team on any given project.
Assemble the right team	Explain the importance of finding optimal in-house team constellations to facilitate a BIM workflow on a project. At times, specialists who work across projects may be required in order to provide additional support to the core team.
The distribution of modeling responsibilities	Assist in these matters and work with project leaders in order to agree on the desired workflow of the team.

PROJECT PROGRESS	BIM Manager Tasks
Explain BIM content creation requirements and BIM library management	Set up and/or maintain a well-organized BIM library; communicate guidelines to BIM authors on how to manage library content between their project and the organization's centralized library.
In-house data management and exchange	Highlight the opportunities and responsibilities related to data exchange from and to BIM; demonstrate how to interface BIM data with external applications in order to make the process explicit to BIM authors and project leaders alike.
Manage the workflow and communicate issues among project teams (in-house)	Depending on the size and business of an organization, the BIM Manager is often responsible for coordinating the efforts by individual Model Managers and/or other team members. Reports about major issues on projects should be communicated to the BIM Manager and discussed in a group on a weekly basis. Such feedback assists the BIM Manager and the others in adjusting their workflow.
Link BIM and engineering analysis tools	Help to determine the workflow between engineering analysis and documentation. The interfacing capability between 3D models used for these at times distinct activities is increasing. Some basic performance checks can even be facilitated via BIM tools in the lead-up to documentation.
Opportunities for construction planning using BIM (4D BIM)	Highlight any potential benefits of linking geometric data to delivery schedules, timelines, or even Gantt charts. 4D BIM gets increasingly used by contractors in order to manage the progress onsite. Head contractors as well as subcontractors should be aware of the 4D BIM workflow and they should adjust their software infrastructure to allow them to engage in these processes.
BIM and cost planning interface (5D BIM)	Point out opportunities for quantity extraction and cost planning using BIM. Smart associations between geometric model and cost data can result in a better integration of information and validation of cost trends on a project.
More and more BIM services are moving from BIM authoring and coordination within offices to the actual coordination of construction onsite.	Explain the opportunities inherent to field BIM and provide examples of its successful implementation to your colleagues. In some cases an organization's leadership may not even be aware of the potential BIM brings to the table in reorganizing the construction site.

MULTIDISCIPLINARY COLLABORATION	BIM Manager Tasks
Assist with the management of BIM workflows in multidisciplinary teams	Explain to staff how BIM can be applied across multidisciplinary project teams. What are the opportunities as well as the typical pitfalls? Explain who should run the BIM coordination and how these sessions get structured.
The purpose and nature of BIM Execution Plans	Explain the logic behind BIM Execution Plans. Those documents are becoming essential guidelines for project teams to orchestrate their coordination efforts. It is therefore pivotal that staff understand their purpose and their effective application on projects.
Share BIM data with third parties	Convey the key criteria to staff for sharing BIM information with third parties from a technical, procedural, and also from a contractual perspective.
Dynamics during multidisciplinary BIM Coordination Sessions	Suggest Best Practice approaches and inform staff about workarounds in order to maximize synergies found in the collaborative process.

The entries above merely represent a snapshot of potential topics to be discussed during the Change Management seminars and workshops. Their content will depend on the core business of the BIM Manager's organization and his or her level of BIM maturity. In the end, it is up to the BIM Manager to determine the most appropriate discussion points to be covered during the seminars.

One other aspect to consider is the grouping of staff into various levels. Change Management workshops should ideally target specific constituencies within an organization, ranging from those modeling or coordinating in BIM, to those supervising the project delivery process, and ultimately also to the key decision makers (leaders) of the organization.

As illustrated in this chapter of the publication, Change Management is a highly relevant and multifaceted process every BIM Manager should be familiar with. Learning how to manage change and being able to engage collaborators in a mentoring role is as relevant as the technical knowledge BIM Managers need to possess. This technical knowledge is the subject of Chapter 3 of this publication. In Chapter 3, Design Technology will be looked at from a number of angles with particular focus on the interface between organizational infrastructures and innovation driven by high-end technology.

Endnotes

1. T. Mayne, "Change or Perish," *AIA Report on Integrated Practice*, November 1, 2005. http://www.aia.org/aiaucmp/groups/aia/documents/document/aias076762.pdf.
2. R.E. Johnson and E.S. Laepple, *Digital Innovation and Organizational Change in Design Practice*, Connecting Crossroads of Digital Discourse, ACADIA 2003 Conference Proceedings, Indianapolis, 2003, pp. 179–183.
3. C. Gray and W. Hughes, *Building Design Management*, Oxford: Butterworth-Heinemann, 2001, p. 12.
4. T.P. Holland, "Organizations: Context for social services delivery," In *Encyclopaedia of Social Work*, Vol. 2, Washington, D.C.: NASW Press, 1995, p. 1789.
5. G.E. Hall and S.M. Hord, *Implementing Change: Patterns, Principles, and Potholes*, 2d ed., Pearson, 2005, p. 7. http://www.pearsonhighered.com/educator/product/Implementing-Change-Patterns-Principles-and-Potholes/9780205467211.page.

6. T. Fritzenschaft, *Critical Success Factors of Change Management*, Wiesbaden: Springer Gabler 2014, p. 62.

7. R. Saxon, *Growth through BIM*, Construction Industry Council (CIC), UK, 2013, p. 27.

8. Management is the "effective and efficient attainment of organizational goals through planning, organizing, leading, and controlling organizational resources," R. Daft and D. Marcic, *Understanding Management*, Cengage Learning, 5th ed., 2005, p. 7. http://www.cengage.com/search/productOverview.do;jsessionid=A60BDF746F97FC611B7E1EF4B4C11479?N=16+4294951002&Ntk=P_EPI&Ntt=3110690487616521236205927533316187724&Ntx=mode%2Bmatchallpartial

9. "BIM's Return on Investment," Autodesk, 2007, http://images.autodesk.com/emea_s_main/files/gb_revit_bim_roi_jan07.pdf

10. RTV Tools online ROI calculator, http://www.rtvtools.com/roi-calculator/

11. T. Fritzenschaft, *ibid*, p. 37.

12. J. Hayes, *The Theory and Practice of Change Management*, 3d ed., Basingstoke; New York: Palgrave Macmillan, 2010, p. 151.

13. P. Mêda and H. Sousa, "Towards Software Integration in the Construction Industry—ERP and ICIS Case Study," Proceedings of the CIB W78 2012:29th International Conference—Beirut, Lebanon, 2012, pp. 304ff.

14. L.R. Bachman, *Integrated Buildings: The Systems Basis of Architecture*, New York: Wiley, 2002, p. 26.

15. K. Kirst-Ashman, *Human Behavior, Communities, Organizations, and Groups in the Macro Social Environment: An Empowerment Approach*, 2d ed., Belmont, CA: Brooks-Cole, 2008, p. 71.

It is impossible to conceive a simple graphic that depicts the typical relationship between a firm's Design Technology and Information Technology. In some cases this relationship may be a straightforward affair: In smaller and medium-sized offices they are likely to be looked after by one and the same person. In larger organizations, that task would become more complex: Depending on a number of factors, the DT department may be the offspring of its IT, or it would be a separate entity that focuses predominantly on design and delivery of projects. In this regard, DT is far more ostensible than IT. Its management is more closely tied to organizations' core business within the construction industry. When considering BIM, the impact of DT reaches far beyond intraorganizational processes. BIM in highly collaborative environments requires focus on interoperability and information sharing across professional and organizational boundaries. Consequently, DT often goes beyond considering support within an organization.

BIM Managers need to learn to communicate their requirements on IT in a timely and structured fashion. This goal often stands in conflict with the fact that projects tend to start spontaneously in the AEC industry. If projects require BIM, there will most likely be a particular configuration for desktop computers that needs to be communicated to the IT support (team) at the earliest point possible. A good number of practices get caught out when their aim to transition an increasing number of projects from CAD to BIM is jeopardized by the lack of available hardware (or software licenses) to run BIM. Budgeted hardware upgrade cycles simply may not consider the additional requirements imposed on the IT team due to BIM. It is the responsibility of the BIM Manager to include these considerations in his/her forward planning and to discuss it with upper management and the organization's IT support.

Hardware/Software License Selection for BIM

One of the most fundamental tasks of BIM Managers is to assist their organization in choosing the most appropriate hardware and software infrastructure to run BIM projects successfully. BIM Managers not only need to know hardware and network specifications that correspond to the various requirements to fulfill certain BIM-related tasks, they also need to be aware of software compatibility with the hardware they specify.

Due to the ever decreasing cost of desktop computer hardware, its configuration and equipment selection have become less of an issue in everyday practice. Still, it is advisable for BIM Managers to get a grip on the impact of software used on hardware configuration. They may find that in some cases, to run certain software smoothly, one needs more (and faster) RAM and/or an upgrade of a machine's graphics card.

A good number of actions performed by BIM software are only optimized for running on a single thread of a computer's CPU; multithread-enabled CPUs may be underutilized most of the time. A BIM Manager is advised to search for CPU benchmarks online before making a decision. Together with IT management, BIM Managers need to make a judgment about the best fit for equipment selection with sustained benefits in consideration of the typical hardware renewal (or leasing) cycles of their organization.

A key aspect for successful management of this technical part of their role is to anticipate future requirements on hardware performance, extent of data storage, speed of information transfer, and the nature of network configuration. One option is to "overspecify" hardware infrastructure, or to commit to purchases that allow for future extensions and upgrades. When it comes to selecting hardware for running the most common BIM software, Bill Debevc, Technical Manager at BIM9, works from the "Rule of 20." He suggests multiplying a BIM file size by a factor of 20 to determine how much RAM is needed to load it. Debevc further suggests taking the fastest RAM available and looking at expansion options for RAM when configuring hardware: "If the system you are getting has four slots for RAM and you are starting with 8GBs, get 2 4GB chips, so you have 2 open slots. That way if you need to upgrade to 16GBs you can do that by adding 2 more 4GB chips."[1]

Looking ahead even further, BIM Managers may, in collaboration with their IT counterparts, go down the path of a transformation of their entire hardware infrastructure toward a cloud-based setup that requires high bandwidth and thin-client end-user interfaces. Either way, there is a danger that recommended specification benchmarks published by software vendors and resellers only reflect the current status quo, but they fail to consider these future requirements. A BIM Manager needs to be on top of the latest technology developments in order to make a judgment on hardware selection that balances between technological advance and the typical hardware upgrade cycles.

Hand in hand with the hardware specifications and software selection goes the programming of BIM licenses required within an organization, or for a specific project. The selection and management of software licenses is a constant cause of headache for a BIM Manager. It often causes friction between them and their firm's management. Why is that the case and why do BIM Managers struggle?

Getting software licensing right is a delicate topic for BIM Managers as it is a major cost they need to justify. In the past, BIM Managers needed to consider the one-off capital expenditure (CAPEX) for purchase as well as the ongoing subscription cost as part of BIM's operational expenditure (OPEX). More and more software vendors are moving to a subscription only model. This development seems to suggest that software vendors now offer more flexible solutions, such as pay-as-you-go access to their software with scalable licensing. In general this appears to be a promising development that addresses the needs in the AEC market. Still, BIM Managers ought to be careful to balance out immediate with future licensing needs and the cost associated with various subscription models—a difficult task given that software developers change, bundle, and rebundle their software suites at regular intervals.

BIM Managers also need to understand the hard and soft licensing options offered by various providers. With hard licensing, users get access to a pool of software licenses that is managed via a network and that gets constantly monitored. In case the number of license requests exceeds the number of available licenses, a license will only be freed up if another user relinquishes his/her current license. A soft licensing model allows users to exceed their pool of licenses temporarily without immediate consequences. Depending on the duration and extent of excess use, the licensing fee gets renegotiated.

The following five suggestions highlight the key aspects of software licensing BIM Managers should consider:

1. Always discuss licensing needs with your organization's IT team/expert.
2. Shop around with different resellers to get the best deal (some may have particular deals with the software vendors).

3. Be cautious about the software your organization really needs; don't get seduced into signing off on preconfigured suites if you are only going to use a small proportion of what they contain.

4. Do your math: Calculate various options of license use with different scenarios that test increased or decreased use over time. Don't get too excited about "once in a lifetime" deals. These change all the time and new ones will eventually pop up.

5. Try to set up a relationship with your software reseller/developer. The better they know your business, the more likely they can tailor their offering to what you need.

Sharing BIM via Networks

One of the key advantages of BIM over CAD is its ability to allow users to work collaboratively on the same model, or a set of federated models in a common data environment. Such collaboration can occur within an organization, as well as across an entire project team. With increasing network speed and evolving methods of data transfer, the geographic location of those contributing to a joint model will soon become irrelevant. As technology enables geographically dispersed teams to share BIM data concurrently, other aspects will need to be considered. What protocols and pipelines are in place to manage the flow of data? This question extends to issues related to access control and file locking, synchronization of model updates between local and central storage, revision control, update notification, and the availability of audit trails to track and document the modeling status.

When working in CAD, operators would usually work on a singular view of the design such as a plan, section, or elevation at any given time. On jobs that require multiple documenters joining forces in CAD, these operators would typically reference in parts of documentation to inform their own work. In BIM, collaborators within an organization are likely to work jointly on a model, contributing their bits to a combined 3D effort that gets viewed as plans and sections for documentation. BIM authors generally work from a local copy of the overall design that gets synchronized with a master file on a BIM server. This method of designing has different demands on work-sharing and combined collaboration protocols such as Worksets. For instance, collaborators work on local copies of the current model version and they borrow parts of it that only they can work on without risking the chance of anyone else working on the model at the same time. The server facilitates model storage, access control, and model element borrowing. Communication features within BIM authoring software, and in some cases third-party instant messaging tools such as Microsoft's Lync™, allow operators to negotiate ownership of certain elements within the design that they can change/update. Consequently, BIM operators need to be aware of the most appropriate intervals for synchronizing the local copies of BIM with the centralized (or cloud-based) Masterfile.

BIM Managers assist their organizations in setting up these collaboration environments in the form of BIM server infrastructure and/or cloud-based interfaces. (Virtual) servers help to regulate the activities listed

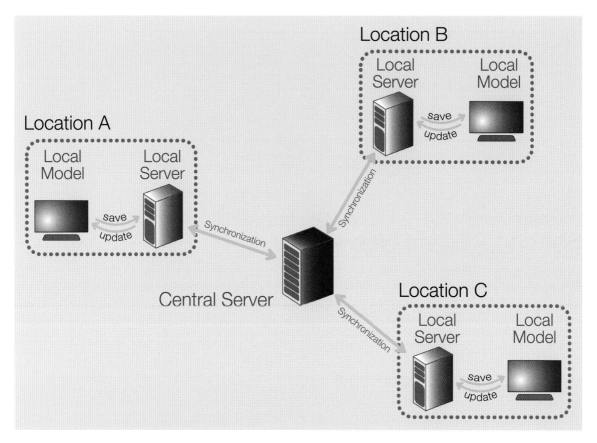

Figure 3–3 Network sharing diagram.
© **Dominik Holzer/AEC Connect**

above; they are often accessible via a web-based admin portal. In collaboration with IT support, they provide the modeling infrastructure for each team within the organization and, at times, they also assist in setting up a common modeling environment for a project team. One key component of that task is to determine the best way to break up a model so it can be easily accessed and updated in a joint modeling environment.

What does the BIM Manager need to know/do in order to support these common BIM collaboration environments?

In its simplest form, the server helps to orchestrate modeling updates by a small number of in-house operators on a project who may well be sitting next to each other. The interactions start to become more complex if collaborators are geographically dislocated, or even working for different organizations, using different software applications. The following section focuses on setting up BIM in a cloud-based environment.

BIM in the Cloud

"As an industry we're always trying to avoid isolated and inaccessible information across project teams; with the Cloud information becomes more centralized and concurrent."

Jon David, Virtual Design and Construction Regional Manager New York, Turner Construction

Cloud computing and related web-based services and applications have been revolutionizing the way operators access and utilize software. Since the first enterprise applications were made available via the Cloud by Salesforce™ or Amazon™ at the turn of the millennium, the spectrum of web-based software services has increasingly expanded. Software as a Service (SaaS) is steadily growing in popularity and cloud functionality is constantly expanding beyond the facilitation of singular applications. The Cloud not only affects the way operators utilize BIM software, it can have a major knock-on or domino effect on the entire IT infrastructure within an organization. Data storage in the Cloud reduces (or eliminates) the need for local server storage; and SaaS applications diminish or even eliminate the need for cumbersome software installations on desktop machines, which may be replaced by simple end-user terminals. Overall, the Cloud promises an organization more flexibility and reduced cost in handling their IT. Such freedom comes at the cost of high dependency on Internet bandwidth and data security.

Processes to Consider

Processes typically applied in the AEC industry such as 3D renders, BIM authoring and coordination, BIM storage, revision management, and document markups by project teams can all be facilitated via the Cloud. With the advent of the Cloud, the rules of engagement between IT and DT get rewritten. Web-based applications with off-premise data storage and secure data management in the Cloud are quickly becoming the norm in an environment where information from manifold sources is seamlessly linked. Facilitated by this information mobility, a number of processes across design, engineering, construction, and operation can be linked together in an integrated way that was previously impossible. Even further, software providers now configure their tools specifically for collaboration in the Cloud such as Autodesk's A360™, Graphisoft's BIM Cloud™, or Trimble's Connect™ applications.

Private Cloud versus BIM Cloud

With a number of private clouds, BIM clouds, and hybrid solutions on the market, it is often difficult for BIM Managers to understand what to look for.

The first thing a BIM Manager should decide when setting up a cloud is whether to go for a private cloud or a BIM cloud where processing occurs in a remote data center. Whereas a private BIM cloud is locally hosted by an organization, providing secured access to project data, operations facilitated via a BIM cloud get processed in a data center. There, information can be accessed from anywhere without requiring local resources.

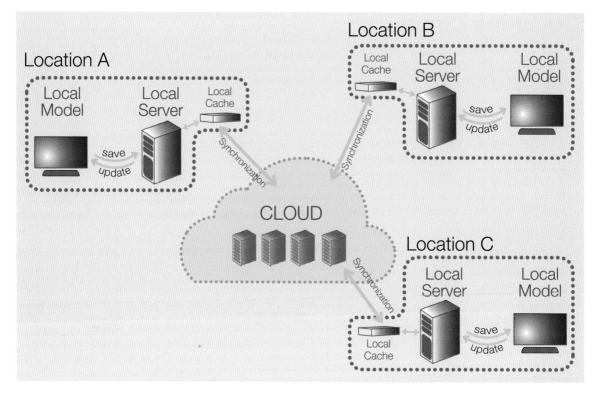

Figure 3–4 BIM sharing in the Cloud diagram.
© Dominik Holzer/AEC Connect

Application and server virtualization software such as Citrix XenApp™ or VMware™ vSphere™, in combination with virtualized graphics processing, are key enablers for hardware GPU sharing. These applications allow firms a flexible setup of their software infrastructure that does not depend anymore on running software via Central Processing Units (CPUs) from individual machines.

BIM Managers need to be aware of the limits of these desktop virtualization methods. For certain model sizes and distance of users, latency can cause user experience to worsen significantly. Often it is not necessarily model file size as such that slows down data exchange, but the amount of auxiliary transactions that need to be processed with each synch between local and master model. These transactions include file opening, locking, closing, and unlocking. BIM Managers therefore need to be strategic on how files are linked together and how shared resources are accessed by those collaborating remotely. Ideally, linked files should be stored at a location where they are accessible to all stakeholders using the same path so they don't need to be uploaded by individuals separately each time a stakeholder synchronizes to the master file. Another way to address delays when managing BIM file synchronization between local and central locations across large distributed end users is to combine methods such as Citrix with solutions that increase currency of data synchronization.

Panzura™ and other systems such as Nasuni™ offer users a single file system to store and manage information from multiple locations globally. File-locking is usually handled via a local cache that reduces latency and ultimately gives users a local-server-like experience. Panzura uses differences between consecutive snapshots both to maintain file system consistency as well as to protect data in the file system. In a process called synching, the Panzura file system takes the net changes to metadata and data between consecutive snapshots and sends them to the cloud.[2] As a result the cloud-based global file-locking mechanism allows users to speed up opening and synching times significantly.

When opting for solutions such as Panzura or Nasuni, BIM Managers need to determine if they want these solutions as a service that includes cloud storage as well. There are flexible options where one purchases the hardware controllers and software licenses to then utilize cloud storage space from a third-party provider.

All major BIM software vendors now offer bespoke BIM cloud solutions and they provide users with a software infrastructure to manage information via their cloud. The key thing to consider when opting for a BIM cloud is the location of the nearest data center where the project information is hosted. As all processing in a cloud is tied to its data center, the farther the geographic distance between user and data center, the greater the latency that can be felt by operators who use BIM clouds. A few hundred or even 1,000 to 2,000 kilometers are usually not of major concern on small- to medium-sized projects (depending on network speed). Once these distances are exceeded and/or a greater number of geographically remote participants interact via a BIM cloud, the lag in exercising a command and the time it takes to synchronize local and master files can slow down collaboration to a point where the system falls down. BIM Managers need to find ways to reduce latency if they want to enable geographically remote teams to collaborate on medium- or large-scale projects.

SETTING UP A CLOUD FOR BIM IN YOUR OFFICE

1. Establish the exact need for the cloud: What kind of services should it offer your organization?

2. Compare the various pricing models available in your local market.

3. Communicate the requirements for operating and maintaining your cloud service with the IT department/support person.

4. Based on your research, select among a range of possible options from private cloud to services such as Amazon, EC2, etc.

5. Be alert about the information storage mechanism when opting for a vendor-bespoke BIM cloud; ensure you will always retain access to your (legacy project) data without any hidden cost factors involved.

Project and Document Management Software

Managing information from construction projects is not purely about authoring and coordinating BIM. It entails a range of activities to interface people, processes, and documents that allow for a project team to collaborate

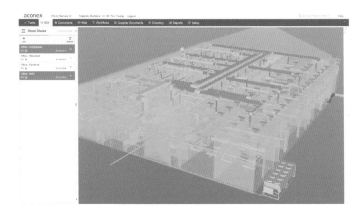

Figure 3–5 Screenshot of
the ACONEX user interface.
© **ACONEX**

effectively. Project collaboration applications have widely been in use throughout the construction industry long before the days of BIM, but they are now taking on board some particular features to support BIM-enabled workflows. The various software applications all offer protocols for information and file exchange, but they differ in certain areas when it comes to the kind of information interface and the manner in which they do so. One feature these tools all have in common is an increasing move toward managing information via cloud-based systems. In most cases, it is not those authoring BIM who prescribe which application the team should use, but the client or the head contractor. Their choice of software depends on the range of activities they like to support across their project team. This is where it becomes handy to understand some of the distinguishing features of various applications. Projectcentre's iTWOcx™ as well as ACONEX™ both include bespoke modules to support bid and tender management processes; iTWOcx also allows users to link document control to Enterprise Resource Planning (ERP) processes. In recent years ACONEX has expanded its capabilities from supporting predominantly 2D document management to offer browser-based access to manage a broad range of BIM-centric design, engineering, and construction activities via the Cloud. Dassault's Enovia™ platform is transferring its lessons learned in supply chain integration and Product Lifecycle Management (PLM) from other industries to the construction sector. With its particular focus on facilitating workflows between architects, engineers, and contractors, Bentley's ProjectWise™ puts strong emphasis on design team collaboration (in particular on infrastructure projects). Similar to ACONEX, it has a proven track record of being able to handle very large datasets, thereby integrating project data from a broad range of BIM and non-BIM applications, with advanced access control management and information navigation options.

Explaining Tool Ecologies

On a recent count of all the digital tools used by the 250+ staff of a prominent Australian design firm, the total number of individual software applications came up to 97. More than half of those tools were dedicated to the design and delivery of projects. These numbers draw a clear picture: We use many different tools to design, engineer, and construct the projects we deliver. Many tools still get used in isolation with little integration of processes, resulting in a doubling up of effort across an organization. Current technological progress highlights the inadequacy of a disconnected approach. Instead of sequential and isolated manual markups or data

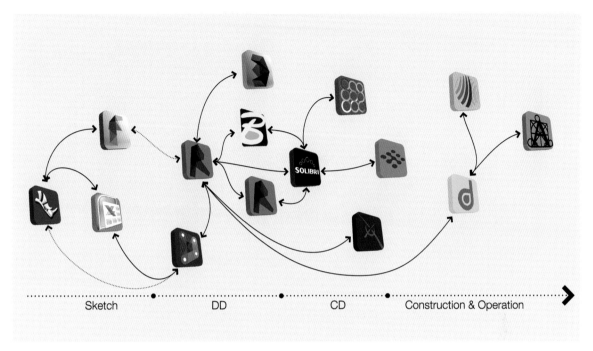

Figure 3–6 Example of a BIM-related tool ecology; focus on interoperability.
© Dominik Holzer/AEC Connect

entry on distinct systems, stakeholders should be able to access and manage project information wherever they are and on whatever platform they use. What role do BIM Managers play within this multifaceted tool environment? How do they lend their expertise in assisting teams in the selection and use of applications that deliver the best results for the project?

From Supporting Singular Software Use to Supporting Process

The practice context faced by BIM Managers is rapidly changing. Such was the nature of project-based work in the CAD era that individual consultants often focused on fulfilling tasks in isolation as efficiently and quickly as possible in order to meet project deadlines. Speed of collaboration came second, hand in hand with an accepted time lag between project revisions. A study published by Panagiotis Parthenios in 2005 revealed that back then, "assisting collaboration" ranked lowest in terms of the functionality designers expected from digital tools.[3]

The industry is transforming: The increasing use of BIM has introduced new software applications to the design and delivery process of projects. Many of these tools are tailored toward speeding up collaboration. Project teams are now considering more consciously how to interface applications used by different stakeholder groups. Interoperability is one side of the medallion; strategic planning of tool use and the establishment of tool ecologies is the other. Interoperability aims at facilitating translation of project data from one tool to another

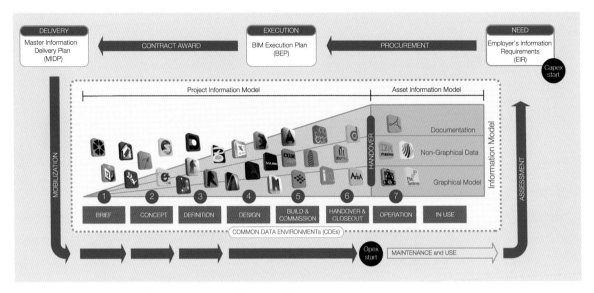

Figure 3–7 Tool use within a Common Data Environment.
© Dominik Holzer/AEC Connect based on UK BIM Task group PAS 1192–2

with minimum loss of data (and geometry) fidelity. Establishing tool ecologies and Common Data Environments (CDEs) refer to a bigger picture.

Establishing Common Data Environments

For large and complex projects, a CDE can help to manage information across the entire supply chain of project stakeholders spanning from planning, design, engineering, manufacturing, construction, and integration with geospatial data and Operation and Maintenance (O&M) activities. Well-conceived tool ecologies, interoperability, and data management have become central issues, as has the fashion in which stakeholders access, review, and manipulate information in teams. Software applications that help facilitate project coordination focus less on coordination of 3D geometric elements for interference checks or other inconsistencies, but they address project management and coordination on a meta level. These tools often specifically address the needs of the client or project manager. BIM Managers need to be aware of the exact goals of project management in order to tailor their own effort and adhere to the data environment desired by the client.

Compensating for End-User Behavior

The AEC industry is in agreement that the one-shop-stop all-rounder tool that can solve every problem doesn't exist. Even if it did, it would most likely not be the most practical application to address the entire spectrum of design- and delivery-related aspects of architectural, engineering, construction, costing, or FM software. Architects will always embrace a range of highly bespoke tools that allow them to fulfill specific tasks, and so do engineers and contractors. Facility and asset managers have an array of applications to choose from in order to address the multifaceted aspects of their work, ranging from a basic asset registry to tracking assets, managing defects, and commissioning data, to energy monitoring, emergency response and disaster planning, maintenance scheduling, and many more.

BIM Managers need to understand the reasons behind the often fragmented tool use across an organization or the project team. Operators in the AEC are typically badly behaved when it comes to a disciplined approach to software use. With countless options to choose, and with an ever-growing and changing tool context, individuals often go for tools they became acquainted with during their training, or they search for tools online that promise a "quick fix" to solve specific problems. As much as this approach is understandable, it causes headaches for the IT or BIM Manager, whose job is to contain tool use as he or she can only support the use of a select number of applications by staff. When it comes to BIM, the strong dependency between various users affords them to be more disciplined. They require guidance about selection, interoperability, and sequencing of tools in the wider BIM context.

Thinking in Ecologies

The BIM Manager's role is to work out the most purposeful way of connecting various processes applied by teams (both in-house as well as across organizations) via a tool ecology that allows users to derive maximum efficiency out of the suite of applications they wish to apply. This is not an easy task: A distinctive approach is required for each project not only because of the different programs to be supported, but also because of the distinctive output requirements by the team and the client. Developing a tool ecology for a hospital project cannot be the same as establishing tool pipelines for a commercial tower project. Demands on data visualization and validation, information management, linkages from conceptual design to construction documents, and coordination of construction sequencing work differently for these types of projects.

Establishing tool ecologies goes far further than simply facilitating interoperability between software applications. When establishing tool ecologies, BIM Managers are the ones who understand the best point in time to move from tool A to tool B; they need to know the workarounds that pick up on the distinctive characteristics of how one tool handles information in order to pass it on to the next. BIM Managers need to align the professional and business imperatives of an organization and adjust tool ecologies for any project accordingly, picking up on the project team's skill level and particular client requests at the same time. FXFOWLE's Alexandra Pollock makes this clear as she argues: "A lot of time your tool-ecology depends on staffing." A BIM Manager's sound judgment of staff software skills that expand into a project team's capability is essential when selecting an appropriate ecology of tools.

On a lifecycle scale, these tasks may at times exceed the level of knowledge that can be expected from any single individual. It is essential for any BIM Manager to liaise with decision makers in their firm to fine-tune their approach and align it with the contractual context of any project. In the United Kingdom, the Building Research Establishment (BRE) suggests a split between managing overarching information requirements of a project in a Master Information Delivery Plan and managing those of stakeholders within an organization in a Task Information Delivery Plan.

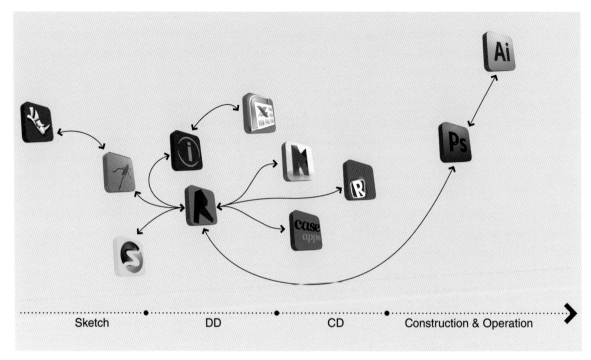

Figure 3–8 Example of a BIM-related tool ecology with focus on supplementary applications.
© Dominik Holzer/AEC Connect

Whichever way the responsibilities between various stakeholders are set up, managing the relationship between geometry and data is of essence when considering tool ecology. Whereas the transfer of geometrical data to BIM has higher relevance for the design and engineering stages, this gradually changes when moving into the construction stages. Further, the management of data stemming from BIM ultimately becomes the key aspect during operation and maintenance. BIM Managers in architecture and engineering firms often misjudge the relevance of the data output for clients and their Facility Managers as they tend to focus too much on early-stage coordination of geometric objects.

Interfacing BIM

The following section sheds light on the key areas that form part of a BIM project's tool ecology, acknowledging that in practice there may well be substantial crossover among those. Starting with the integration of geospatial and BIM data, principles behind interfacing BIM and topology models get explained, to then focus on ways to manipulate data and data structures inherent to BIM. The section then looks at BIM output for fabrication, construction programming, and for the exchange of data from virtual models and physical measurements in the field. Finally, the section looks at the transfer of BIM data to tools that form part of Computer Aided Facility Management (CAFM) software environments.

A clear overview about tools that link information into BIM, those that support information management within core BIM applications, and those that handle information extracted from BIM, is difficult to obtain. These distinctions are gradually eroding due to the ever-expanding list of applications interfacing with BIM workflows. As an example, using 3D virtual models for simulation and analysis has traditionally been a process that sat outside of BIM. Nowadays, more and more analysis tools either become integrated into BIM platforms, or tight cross-referencing of BIM authoring and engineering analysis functions between distinct tools get enabled. With the increasing development of software that supports BIM collaboration across organizations in the cloud, further convergence of core BIM tools and those that previously sat in its periphery are likely.

Geospatial/Point Clouds to BIM

BIM Managers on the architecture as well as the contractor side are often faced with interfacing their BIM authoring tools with information stemming from land surveyors. If, traditionally, 2D topographic line-work was the base of interaction between these parties via CAD, recent technological developments have radically transformed the methods for information exchange. A recent study undertaken by the Australian Institute of Architects and Consult Australia[4] reveals that the use of tape measures, theodolites, dumpy levels, and staffs is increasingly giving way to total stations, GPS, GIS, and laser scanning technology. In addition, the use of publicly accessible Geospatial Information Systems (GIS) data is becoming more common.

Common BIM authoring applications usually come with options for importing survey data directly into their 3D modeling environment. The data collected in the field by the surveyor is manipulated in specific surveying software to generate items such as Digital Terrain Models (DTMs), Triangulated Irregular Networks (TINs), strings, and point cloud files. The creation of an accurate and reliable "existing conditions" Survey Information Model (SIM) is vital to support the design process in BIM. Merging point clouds with modeled content allows BIM authors to juxtapose existing context with newly generated design. There are many uses for this approach. Designers can take measurements with high accuracy when working on project extensions or refurbishments. Planners can determine the exact distances between a newly developed project and existing vegetation such as trees. Contractors can use the point cloud data for precision setout onsite.

BIM Managers should request these geospatially located SIM files from land surveyors in order to load them directly into their BIM authoring applications. Facilitating this process, BIM Managers work closely with the land surveyors and assist them in increasing their working knowledge of BIM and agree on the desired interface with their primary BIM authoring application. BIM Managers need to be aware that surveyors use different methods for locating features depending on required accuracy. Failure to specify accuracy requirements at the outset can greatly diminish the quality of the data delivered. In order to address this dilemma, BIM Managers should create a "survey brief" for the surveyor with a detailed description of all the BIM-related data requirements to be met. The brief contains prescriptions of BIM file formats, benchmark locations, units, and coordinated systems, as well as a clear definition of what is to be surveyed and to what level of detail.

The consultant/surveyor interaction facilitated by this BIM-enabled workflow allows for more ongoing exchange of geospatial data throughout the building lifecycle. (Head) contractors can then utilize the resulting BIM data for digital setout and other Field-BIM-related activities.

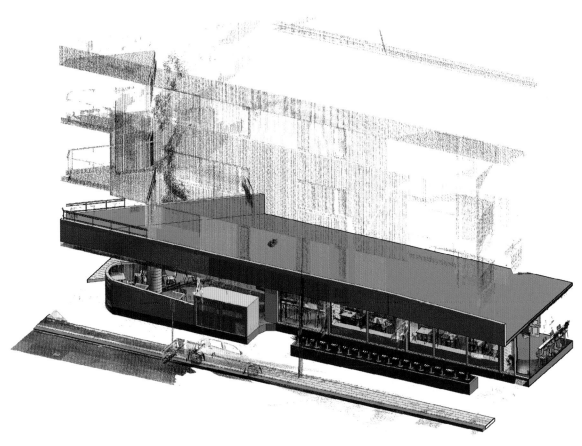

Figure 3–9 AAM point cloud scan—Revit Café sample.
© **AAM Pty Ltd**

Surface Models and BIM

When analyzing processes in design practice one will rarely find that projects are initiated directly in a BIM application. BIM tools are more often than not seen to restrict early-stage design exploration, and alternative tools for mocking up 3D surface models are used instead. What is the right time and what is the right approach to move from topological models that are used for form-finding into BIM? Can one "post-BIMify" geometric models that weren't initially set up with object-oriented assembly and data association in mind? Alternatively, is it wiser to remodel from scratch in BIM, and if so, what can be taken forward from the initial surface model as a reference?

A single answer about the right time to move to BIM on a project doesn't exist. It depends on a variety of factors. BIM Managers need to be able to assist in determining the best path to take based on the existing skills within the team, the software available to them, preferences of the project leader, and the desired output throughout various project stages. One strategy is to use both surface models for morphology exploration in parallel with

BIM massing models of the same project for quantitative feedback. Renowned architects Woods Bagot recently delivered the NAB 700 Bourke Street building in Melbourne, Australia, where tool ecology played a significant role in designing and documenting a complex facade system. Initially, the morphology of facade elements was advanced via digital sketching in Autodesk 3Ds Max™ and McNeel's Rhino™. It was then advanced parametrically with the Rhino™ plugin Grasshopper™. Geometry was then exported to Autodesk's Ecotect™ for solar/performance studies.

Based on this performance analysis, the Rhino™ model was updated in order to generate the precise boundary where the floor slabs connect to the facade. These lines were then exported as key references to start rebuilding the facade system in the BIM authoring tool Revit™. Rhino™/Grasshopper™ allowed Woods Bagot the freedom to explore geometric relations; Revit was the right tool document in which to visualize the NAB Docklands project.

As explained above, BIM authoring applications allow for the import of non-object-oriented geometry. In most cases, though, this geometry serves as a backdrop or reference to reconstruct the design via the BIM authoring tool. This is a one-way street with no ability to associate the geometrical data either parametrically or via other

Figure 3–10a NAB 700 Bourke Street, color distribution in an exploded axonometric of the facade system.
© **Woods Bagot**

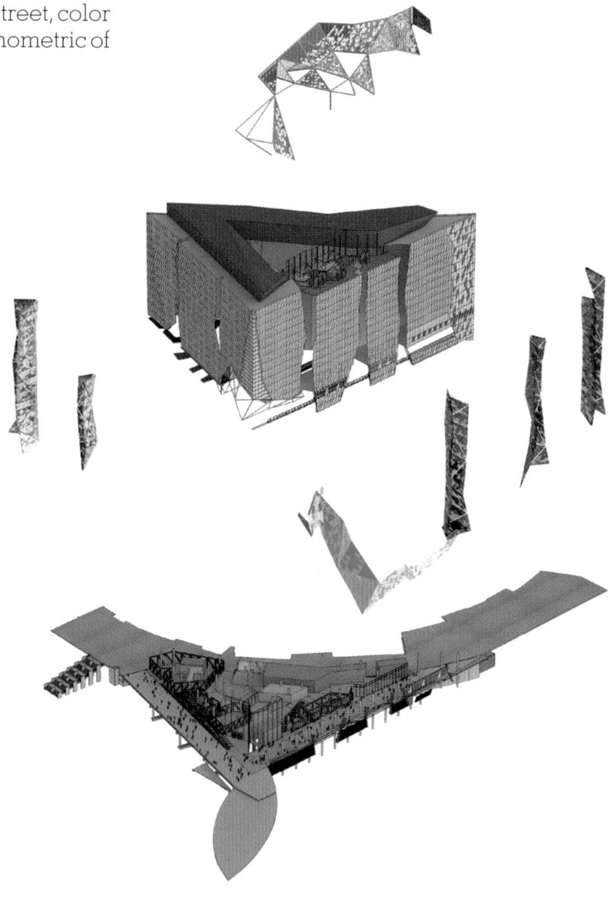

Figure 3–10b NAB 700 Bourke Street, color distribution and closeup of a facade panel.

© Woods Bagot

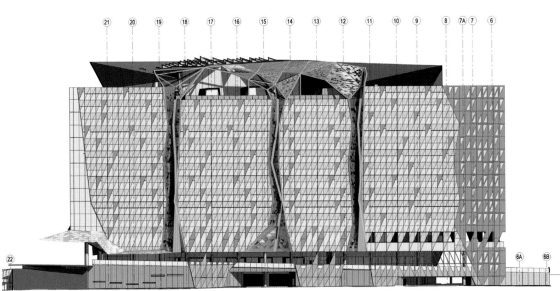

Figure 3–10c NAB 700 Bourke Street, East Elevation.

© Woods Bagot

means to BIM. Ultimately, this means that the logic applied to generate the surface model in the first place cannot be translated into informed objects in BIM.

More recently, solutions are emerging to bridge the split between "flexible" surface models and object-oriented BIM. Geometry Gym's Jon Mirtschin has developed a range of tools that allow for these two paradigms to interact. Geometry generated in a freeform modeling tool such as McNeel's Rhinoceros can, with the help of Mirtschin's tools, be read into a BIM authoring platform (most likely with a smart IFC translator) where geometry is broken down into elements that authoring software can interpret and additional object properties can be added. Nathan Miller, Associate Partner and Director of Implementation at CASE Inc., has come up with data schemers that allow for intelligent and live association between surface modeling applications and BIM software. Rhynamo is an extension for Autodesk's Dynamo™ (discussed later in this chapter), which allows the reading and writing of Rhino files; the process allows for maintaining dynamic links between the surface model tool and the data-heavy BIM platform. By extracting geometric information one is interested in and appropriating geometrical definitions, Rhynamo enables BIM software to interpret that information.

Interfacing BIM and Engineering Analysis

Traditionally, one of the most wasteful aspects of digital modeling has always been the split between modeling for documentation/visualization purposes and for simulation and analysis purposes. In other words, in the

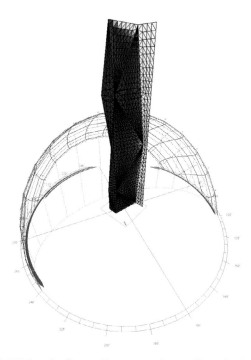

Figure 3–11 NAB 700 Bourke Street, Ecotect analysis of facade elements.
© **Woods Bagot**

past engineers typically needed to remodel the architect's design in order to suit the particular demands of the simulation software they applied for performance testing. Even worse, a model suitable for daylight analysis undertaken by environmental designers would often not be suitable for analysis by structural engineers or smoke-spread testing undertaken by fire engineers (just to name a few). The reason for this disconnect lies within the differences of semantic geometric model representation either as surface model, centerline model, or solid-geometry definition. In addition to the different requirements on geometric data, varying analysis tools also require different resolution or abstraction of geometric entities in order to be able to compute performance data and derive accurate simulation responses. All of the above made it close to impossible in the past to use one model for design, 2D documentation, and 3D visualization, as well as performance testing. With many models existing in parallel, professionals from varying backgrounds, but in particular engineers, had to remodel their analysis geometry each time the architects updated their design.

Advances in the way BIM geometry gets configured as well as advances in interfacing between BIM and analysis software increasingly allow stakeholders in the industry to live-link model information between design and analysis software. Some software providers have worked on bespoke translators that communicate BIM data from the authoring tools to the analysis tools within their suite of software applications (e.g., Revit™ to Robot™). In those cases where applications from different software companies get used, there exists the option to link together analysis model data and BIM via custom developed schemers (see Geometry Gym) or data management tools such as Dynamo™/Rhynamo™. The IFC format is often the common denominator needed to maximize the transfer of both geometrical and non-geometrical information.

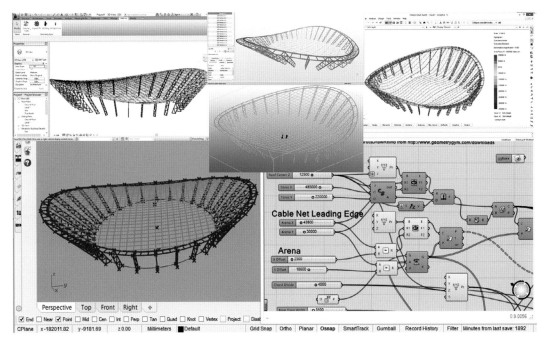

Figure 3–12 A parametric stadia model by Geometry Gym.
© Jonathan Mirtschin, Geometry Gym Pty Ltd

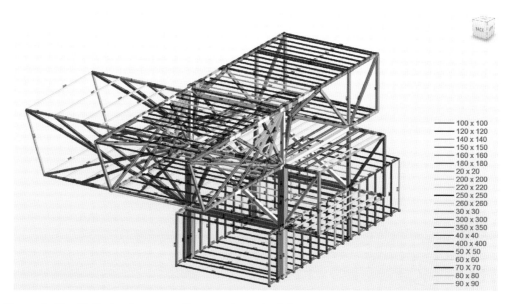

Figure 3–13a Optimized cross-section member schedule.

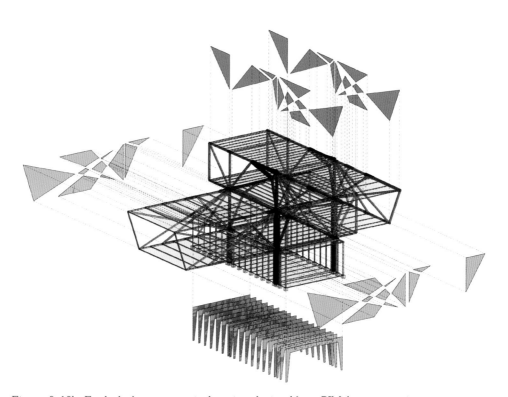

Figure 3–13b Exploded axonometric drawing derived from BIM documentation.

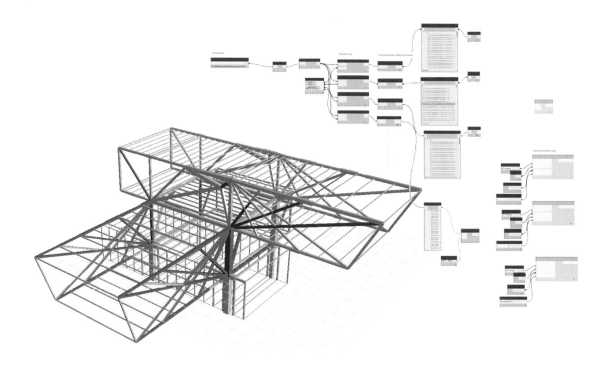

CAD to BIM (Building Information Modeling)
- CAD Geometry can be transferred to BIM platform

Figure 3–13c CAD to BIM via Dynamo.
© Junghwo Park (Point One Studio), Permission per Lect. Franz Sam, Dipl Ing. ARCH UNIV AK Wien

The IFC Question

Any BIM Manager will, at some point, be confronted with Industry Foundation Classes (IFC).[5] In its fourth release at the time of this writing, IFC is a passionately debated BIM topic, hailed by those who engage with its core concept, cursed by some who have experienced its ugly side (if one can talk about such). Much has been written about the principles and usefulness of IFC. There is no need to reiterate any of this information here. What is relevant for BIM managers, though, is to understand the practical implications of using IFC on projects and how to structure support around facilitating its use across an organization.

One of the most active users and commentators on the IFC format and process is Rob Jackson, Associate Director and BIM Manager at Bond Bryan Architects in the United Kingdom. Reflecting on the success factors for making IFC work for his organization, Jackson previously reported: "The truth is that IFC requires a detailed understanding to get the best out of it. You have to understand how it works and you have to understand both the software you are importing from and the software you are exporting to. At Bond Bryan, one pivotal step to maximize BIM data fidelity via robust import/export of IFC, was to employ a staff member who would specialize

ware. There are options for setting up standard textures in the BIM authoring software, which can be passed on to the visualization tool. Once there, additional definition to the textures can be applied (as well as some reduction in polygon count); this model information can then, in turn, be forwarded on to virtual reality software that interacts with wearables. For larger or more detailed projects where high-end output is required, hardware limitations would make it more reasonable to remodel BIM output in specialized software, in order to minimize polygon count and exert more control over the visualization of geometry.

One other form of deriving graphical output from BIM refers to process visualization capture. The focus lies less on photorealistic output, but on the illustration of the construction process via a sequence of images (or a movie). This 4D BIM programming of construction progress is tied to planning and scheduling of various activities onsite by the project managers who work for a head contractor. It provides visual feedback that assists the project manager and others in assigning activities and managing resources before and during construction. The principle behind this form of visualization is simple: The sequence of work undertaken by individual trades gets time-lined in a common Gantt chart (by using stand-alone BIM coordination tools or by associating those with scheduling tools such as MS Project or Primavera) that then gets synchronized with the referring construction activities and equipment locations in a 3D BIM environment. That way, when the Gantt chart gets animated, the construction sequence unfolds.

Advanced features in bespoke sequencing plugins/tools allow users to add temporary works to the permanent ones, construction machinery (such as crane movements), or actual movements of construction workers to the animation. In addition, the base model for process visualization capture can also be exported to dedicated rendering tools in order to add textures/lights to facilitate better output quality of an animation. That way the 4D visualization output can be taken further to illustrate the construction process to the client or other relevant parties.

Contractors can enhance "safety onsite" via this form of process capture as potential risks can be identified. BIM coordination tools such as Solibri have several built-in rule-checker options for the automated detection of design deficiencies. It is likely that we will see an increase of capabilities within coordinating tools where local safety-related regulations can be encoded in a fashion that will make it possible to automate the flagging of safety concerns for review by construction teams.

BIM to Fabrication

One aspect of BIM that is rapidly gaining importance is the automation of the transition from design to fabrication. Being able to fabricate straight off a shop model without the need to abstract information via 2D documents that form the contractual deliverable is a logical step in the evolution of BIM. This process doesn't work for every design aspect. It makes the most sense where the building objects or systems to be assembled are based on a set of standardized or modular components, or if these components can easily be manufactured with available equipment. Steelwork for structural elements, mechanical/hydraulic ducts and pipes, facade systems, structural timber—these are all suitable applications. BIM Managers operating on a subcontractor and fabricator level should get a grasp of the opportunities lurking within rapid manufacture via BIM. The two major pathways for achieving this are either the use of open standards such as IFC, or the setup of bespoke

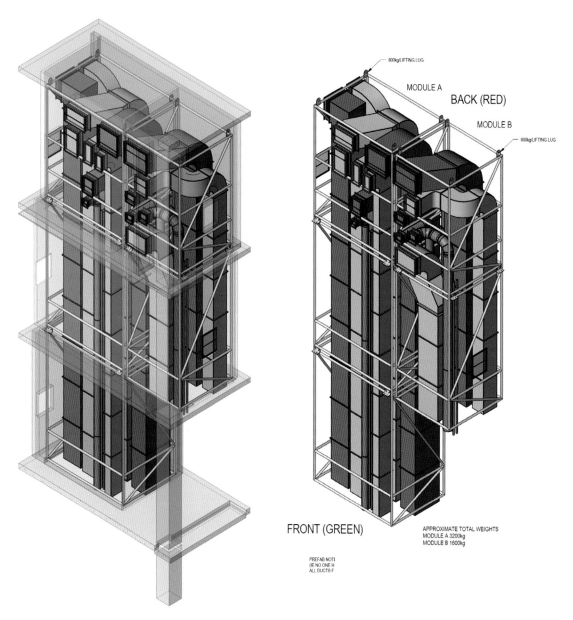

Figure 3–17 Mechanical BIM shop model ready for fabrication.
© **A.G. Coombs Group**

industry solutions[9] that maximize interoperability within the constraints of a preselected tool ecology. Using virtual models straight for fabrication has long been common in parallel industries such as automotive and aerospace manufacturing. The building industry only slowly wakes up to the opportunities of rapid manufacture and preassembly of systems.

Figure 3–18 Paperless jobsite—Field BIM.
© **Turner Construction Company**

Additional benefits of this process include the transition from construction-focused assembly to manufacture-focused pre-engineering and production. Hand in hand with the opportunities to shortcut the design to fabrication process come opportunities for supply chain integration and the connection of Bills of Material (BOMs) stemming from BIM to Product Lifecycle Management (PLM) and Enterprise Resource Planning (ERP). These systems tie into a firm's transaction-related activities such as resource planning, purchasing, storage, QA, productivity management, HR, finances/payroll, and more.

BIM Anywhere

The days are over where there existed a clear distinction between BIM undertaken in an office setting and construction work undertaken in the field. Technological advances in hardware as well as software result in ever more interaction between design and construction processes. Being able to access and interrogate project information right where construction occurs offers enormous benefits to contractors and others. Jon David explains how U.S.-based Turner Construction progressively moves toward BIM in the field: "Paperless jobsites are something we began implementing a few years ago. We are continuing to advance our use of mobile applications: multiple users accessing a database of information for quality control, verification of installation, and writing daily construction reports, to name a few. There is an extremely large amount of data recorded daily by field engineers and superintendents via mobile devices."

Data from the field is no longer just gathered manually, but barcode scanners, RFID (radio-frequency identification) tag readers, 3D laser scanners, and sensors help to capture asset information and building performance and other relevant data. On the other end, data from virtual models increasingly gets brought to the field, closing the gap between design, installation, and commissioning. Atul Khanzode leads technology initiatives as Director of Construction Technologies at DPR Construction Inc., one of the most prominent contractors on the U.S. West Coast. He strikes out how their approach makes DPR one of the most successful firms using BIM to bridge between design and construction: "At DPR we have a great number of what I would call 'craft employees'; in contrast to many other large contractors we build our own thing. We see a lot of value of BIM from that perspective and we invested a lot of money to build up our BIM resources. BIM Management is a skill set necessary to have in-house. Our basic philosophy is that we want people who install in the field to produce the model!"

BIM to FM

There exists a large amount of Computer-Aided Facility Management (CAFM) software covering a broad range of activities and processes undertaken during operation and maintenance of a building. These activities include gathering data in an asset registry, managing defects and commissioning data, monitoring and controlling systems in a building automation system, emergency response and disaster planning, maintenance scheduling, and many more. If formatted adequately, linking data from planning and construction to these tools via BIM can be very useful for facility managers and asset owners. The problem is that often facility managers are not aware of the potential to link BIM data to CAFM tools. Andrew Tape, Global BIM Coordinator at Benoy, explains: "Facility Management traditionally hasn't thought about requesting the information they require from the outset. Even six months prior to handover their requirements are still unknown, and it generally isn't until a month before handover that the information they need to manage a facility is considered."

BIM Managers on the design and construction side are usually not too familiar with the tasks performed by CAFM systems. Linking BIM data to FM represents a fairly recent effort that has slowly gained momentum since the early 2010s across the mainstream industry. The UK level 2 BIM requirement is one example where authorities are now actively pushing for stronger integration between BIM and FM. The mandate asks all project and asset information, documentation, and data to be electronic by 2016. Applying data schemers via the Construction Operation Building Information Exchange is one way of achieving connectivity between BIM and FM. At the same time, there are also other applications to support BIM to FM integration between geometrical and object-related BIM data and Facility Management Processes.

Some tools primarily support the capture of room data appropriated for BIM during design and construction, but their developers now expand on that capability via a plugin that caters specifically to FM aspects beyond

Figure 3–19 Connecting BIM to FM data, Zuuse Interface, Zuuse Pty Ltd.
© **Zuuse**

commissioning. Other tools historically come from the FM and asset management corner. They are now being augmented with capabilities to interface with BIM models and the associated workflows. Other tools again are primarily developed to facilitate the interface between BIM and FM/asset management (with tools such as Zuuse™ and Ecodomus™). A new breed of BIM tools aims at addressing building lifecycle activities comprehensively by including information management activities from early feasibility studies to design, construction, and operation.

BIM to FM software applications usually don't supersede all functions of traditional CAFM tools. They are likely to connect to them and combine a data storage mechanism with intuitive management/interrogation of design data, a supporting 3D geometry engine, and manifold interfaces to other tools and formats. Despite their differences, BIM to FM tools have one thing in common: They all tend to include web/tablet-enabled user interfaces that consider activities in the field, while allowing operators to manage O&M data in the Cloud with multiuser access.

Future Developments

As stated at the beginning of this chapter, technology is the most transient aspect of BIM. Whereas social, professional, legal, and business drivers for BIM only slowly undergo change, technology is in a constant flux. Due to this fact, any predictions about future developments of BIM technology are difficult to make. Still, a number of trends are beginning to show.

BIM TECHNOLOGY TRENDS

1. We are only starting to take advantage of what BIM in the Cloud has to offer. Project teams will increasingly interact using cloud-based "software as a service" solutions, and they will increasingly do so to enable collaboration in real time. Next to this development, we are also going to see an increased focus on network speed and reliability, working with "thin clients" and other end-user interfaces that are platform agnostic.

2. Higher network speeds and Wi-Fi connectivity will allow us to move toward greater independence from specific hardware (challenging the current focus of IT specialists in the construction industry). It will lead to increased proliferation of Field BIM onsite with better integration of construction processes with ERP and PLM. To be more specific: In the future we will see more direct interfaces between design, model-based specification, resourcing, ordering, storage, (offsite) pre-assembly, installation processes, QA, commissioning, time-sheet management, remuneration, and other business-related activities.

3. The boundaries between models for design, engineering, documentation, and construction coordination will diminish. Software providers as well as end users will develop intelligent schemers that allow users to exchange model data multidirectionally across a number of applications that were previously working predominantly in isolation. There is no trend indicating that we will be left with one-tool-fits-all-purposes software, but rather suites of tools that are highly interoperable.

4. BIM authoring and coordination tools will become more intelligent. An increased number of model-checking functions will become standard, and there will be options to apply plugins for checking against local codes and regulations.

5. Fabricating directly off the model. The use of 2D documents as a medium to communicate construction requirements will diminish. As we increasingly push tool ecologies straight to production processes we will be able to work straight from datasets inherent to the model to drive fabrication equipment. A surge in affordable and large-scale 3D printing and the use of robots in construction will give us more freedom to explore novel design solutions via high-end technology.

6. The advent of the age of data. When it comes to data management via BIM we are merely scratching the surface of what is possible. The future is bright for those who master the link between geometric models and a great number of data sources that can be associated with them (such as geographic information systems, GIS). The process of modeling has long been in the foreground of BIM; now we refocus our attention on honing into the potential of using design technology as a conduit to interface datasets from an infinite number of sources such as commerce, transport, food production, and environmental sustainability to our built environment. This convergence will start with clients who use BIM for verification of key performance financial data related to their entire portfolio or enterprise. It will then expand into the setup of precinct and city information models and proliferate further into the entire built environment globally.

Technology will keep on transforming BIM as we know it. BIM Managers need to be aware of these developments and consider them in their strategic thinking and support for their organization. The next chapter will focus on exactly that—the crucial task of BIM Managers to set up a support infrastructure for the effective delivery of projects within their organization and the wider project team.

Thank you to all the experts who so generously offered their thoughts and insights for this chapter: Alexandra Pollock of FXFOWLE, Brok Howard of HOK, Jon David of Turner Construction, Rob Jackson of Bond Bryan Architects, Shane Burger of Woods Bagot, Jan Leenknegt of BIG, Atul Khanzode of DPR Construction Inc., and Andrew Tape of Benoy. Interviews were carried out by the author in person and via correspondence during the Spring of 2015.

Endnotes

1. Bill Debevc: Configuring a "Killer" Data Center for Your Revit Users, Revit Technology Conference U.S. (2012), accessed March 22, 2015: http://rtcevents.net/rtc2012us/materials/S06%20Configuring%20a%20 Killer%20data%20center%20for%20your%20Revit%20users-Bill%20Debevc_Handout.pdf.

2. Panzura Descriptor, accessed March 22, 2015: http://panzura.com/wp-content/uploads/2013/01/ Snapshot%20Technology%20Brief.pdf.

3. Panagiotis Parthenios: "Conceptual Design Tools for Architects," Harvard Design School, PhD (2005), p. 91.

4. In 2013, the Australian AIA and Consult Australia published four series of BIM practice documents. The "Outreach" series contains 10 short papers highlighting the relation of various industry stakeholders with BIM. The O9 document: "Surveying for BIM" outlines the changing relation between Design Consultants and Land Surveyors. The paper can be downloaded upon free registration at: http://wp.architecture.com.au/ bim/groups/outreach/, accessed April 14, 2015.

5. Industry Foundation Classes (IFC) BuildingSMART is all about the sharing of information between project team members and across the software applications that they commonly use for design, construction, procurement, maintenance, and operations. Data interoperability is a key enabler to achieving the goal of a buildingSMART process. BuildingSMART has developed a common data schema (IFC) that makes it possible to hold and exchange relevant data between different software applications. See http://www. buildingsmart.org/standards/technical-vision/open-standards-101/, accessed March 22, 2015.

6. http://bimcrunch.com/collaboration-bridging-the-divide/, accessed March 22, 2015.

7. A general overview about such tools can be found at https://aec-apps.com/, accessed March 26, 2015. Data association and manipulation tools include applications such as BIMlink™, iConstruct™, or RTV Reporter™. Library management can be facilitated with tools such as Unifi™, Content Studio™, or Kiwicodes™.

8. There is a large number of such cloud-based applications emerging; they vary in nature and functionality. Such tools include BIMcollab™, BCF Manager™, ADSK 360™, Sparkframe™, BIMx Pro™, ProjectWise™, Trimble Connect™, and many more.

9. One such solution is the Australian BIM-MEP[Aus] initiative with its "BIM all the way" approach.

BUILDING UP A BIM SUPPORT INFRA-STRUCTURE

4

Developing a support infrastructure that enables others to implement BIM is at the very core of the BIM Manager's role. Though a solid grasp of technology provides an essential vehicle for actualization, the framework can only be established through a well-formulated strategy for change. Ultimately, a well-articulated BIM support infrastructure has the potential to go beyond the merely operational and empower those who work with BIM to excel at what they do.

Helping others to adopt BIM is a multifaceted task. Next to technology savviness, it requires in-depth knowledge about a broad range of standards, policies, procedures, and workarounds. Consolidating this knowledge into easily digestible guidelines, tutorials,

0

and other forms of support material and propagating this material among colleagues is a BIM Manager's art to master. This chapter explains how BIM Managers develop BIM Standards and other guidelines, structure their BIM Libraries, set up and advance BIM Execution Plans, flesh out BIM Capability Statements, and put together a well-considered BIM training program. This chapter also explores how BIM Managers grow their in-house BIM support infrastructure and disseminate their knowledge to facilitate peer-to-peer support.

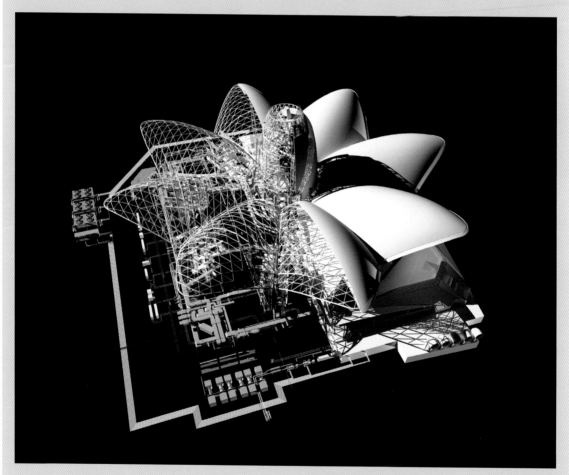

Figure 4–1 Caspian Waterfront, Baku Azerbaijan.
© Benoy

Propagating BIM

The first chapter about Best Practice BIM explained why it is most probable that over time BIM will become an integral part of project delivery. There will be less and less need to differentiate between those delivering projects and those supporting them using BIM. In order to reach this goal, BIM Managers have an obligation to advance their colleagues' BIM skills and to assist them in boosting productivity within their organization using BIM. When doing so, they do not need to start from scratch. A number of industry groups around the world have developed support material to engage with this task. These include the United Kingdom's RIBA Digital Plan of Work together with the NBS BIM toolkit, the Singapore BCA's BIM Guide, and buildingSMART Finland's COBIM (Common BIM Requirement 2012)—just to name a few. BIM Managers have a responsibility to act as a conduit for top-down information/requirements to be aligned with in-house modeling and coordination processes by staff.

In order to get their support right and to achieve higher productivity, BIM Managers monitor and review national policies and specifications, guidance documents, contract forms, latest tool releases, market movements, and much more. Based on those inputs, BIM Managers bring these outside influences into context with an organization's desired BIM capability and establish pathways for change. Lee Wyles, Project Technology Manager at BDP—one of the United Kingdom's leading multidisciplinary consultancy firms—explains it this way: "We started implementing a mandate to regulate workflows and we provide clear roadmaps of how Design Technology and BIM are to be implemented."

The dream scenario of any BIM Manager is to facilitate well-oiled machinery whereby team members collaborate in BIM requiring little to no outside assistance. It lies within the responsibility of the BIM Manager to help others to increase their skills and, in those cases where BIM requires specialist input, supporting their day-to-day actions with "back of house" support material and standards that optimize their workflow. In doing so, BIM Managers allow those delivering projects to focus on their core deliverables rather than what can be considered as an auxiliary activity/process. It is therefore the BIM Manager who needs to understand how to either help others, or to help them help themselves.

By nature, a great number of employees in architecture, engineering, or construction firms will try to pass on BIM-related activities to the BIM Manager. How can BIM Managers avoid being misused as ad hoc project support without advancing skills among their colleagues? BIM Managers have to strike the right balance between giving support and allowing others to learn and advance their own skills. When it comes to technology, BIM Managers are more often than not distinct specialists among a group of generalists who focus on design, engineering, project delivery, and so forth. These individuals—most likely the great majority—cannot be expected to invest their time in order to be on top of the latest BIM developments.

What counts for most at the end of the day is support that has an immediate and positive impact on their project work. Therein lies the problem. The most direct way is always a one-on-one conversation with a colleague who is in need of help. This approach is nevertheless not sustainable in the long run, and BIM Managers need to look for different ways to disseminate and "democratize" their knowledge. One means of achieving this is by helping nurture a group of individuals with above-average BIM skills. These "Model Managers"—BIM architects, engineers, or project BIM Leaders—become the middlemen between the big-picture BIM Strategy and work undertaken by individual contributors. Another crucial tack is the generation of in-house support material.

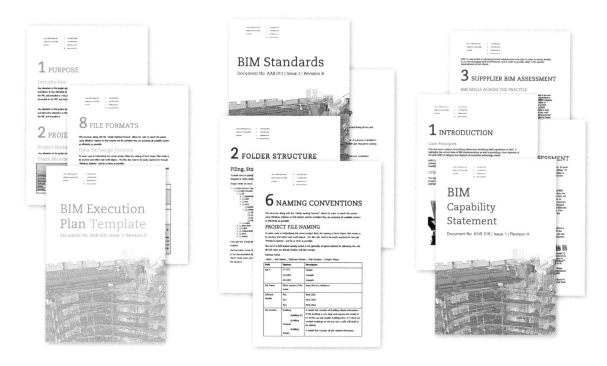

Figure 4–2 Examples of BIM support documents.
© Dominik Holzer/AEC Connect

The medium for proliferating support material is often a combination of printed documents, online resources, libraries, and templates produced by the BIM Manager and his/her team. One of the biggest mistakes made by BIM Managers is to assume that great support documents equal great support. In fact, common BIM operators rarely read these documents. They are simply too busy doing their work and they just want a solution that works for them. BIM Managers therefore have a responsibility to complement their writing of support material with continuous promotion of their content. They have opportunities for doing so with the inception of a new staff member, during project startup meetings, during regular project reviews, via regular internal BIM/Technology newsletters, and during educational sessions with Model Managers and Upper Management.

Hence, BIM Managers need great communication skills to share bite-size bits of information in a fashion that produces the best practical results at a project level and beyond. In some cases users simply need to know that specific standards or guidelines exist, so they can refer to (and adhere to) them while working on a project. In those cases users need to know how to look up the information they are after. Other support material such as templates or BIM Library components get used on projects so BIM operators need to be able to find and access them in the most effective way. Some support material such as BIM Execution Plan templates are most relevant for project leaders who are responsible for initiating a project. They need to, though, be referred to and updated throughout the various stages of a project. The dissemination of knowledge also depends on the context—the size and type of firm.

What are the key bits and pieces of information that need to be understood by an organization when it comes to BIM? The following section lists a number of support areas that require special attention by BIM Managers. In the past, BIM Managers tackled support for these areas predominantly one by one. It becomes apparent though that they form part of an integrated whole. BIM requirements by clients have an influence on how in-house BIM Standards should be set up. These standards relate to one's capability and resources, which in return affect what's in the BIM Execution Plan. Some BIM support material remains rather static and unchanged throughout a project's lifecycle; other support needs to be adjusted continuously. BIM Managers need to balance out what can be known about the way BIM gets utilized, and what still needs to be established during the various project stages.

Starting with the End in Mind—Employer Information Requirements

"Clients have no clue what they want to get out of BIM." This is a common sentiment among a great number of BIM Managers who are asked to work toward their clients' information requirements. Why is such skepticism prevailing among BIM Managers when it comes to their clients' understanding? Historically, there has been a mismatch between the declared (or often not so declared) expectations by clients and the BIM goals of consultants and contractors. In other words, clients often do not get the value-add from BIM they expect. Immoderate or badly communicated deliverables can lead to frustration as clients feel that their design and construction teams overpromise and underdeliver when it comes to BIM.

This sentiment is not surprising: BIM Managers in consultant and contracting firms usually focus on the immediate benefits of BIM for their organization. It is hard enough to justify to the decision makers within their firm what they do and to make a business case for it. Second-guessing BIM's value-add to clients exposes BIM

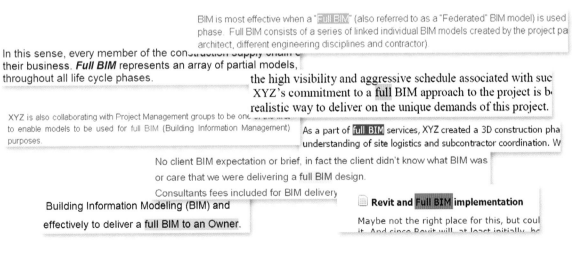

Figure 4–3 "Full BIM" text extracts from project briefs and other documents.
© Dominik Holzer/AEC Connect

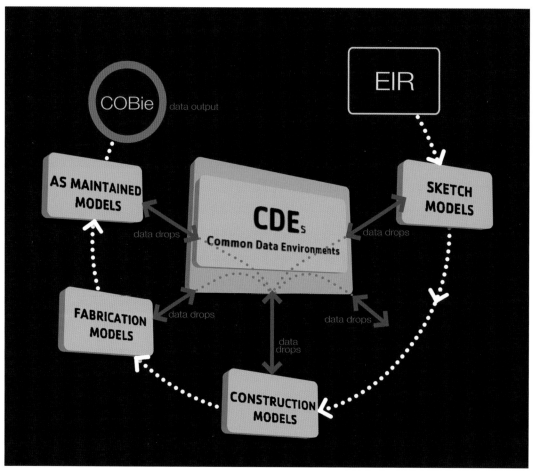

Figure 4–4 The principle behind Employer Information Requirements.
© **Dominik Holzer/AEC Connect**

Managers to the risk of adding deliverables that may not form part of their professional service agreements. Why should a BIM Manager care about client BIM requirements if at first they have little to do with the traditional way their firm delivers projects?

There exists a simple answer to this question: Employer Information Requirements (EIRs) are a mechanism for clients to declare what they are after when they ask for BIM. They form a pivotal instrument to realign the BIM output by design teams with the desired benefits of BIM for owners and operators of assets. EIRs don't solemnly report on information to be handed over at the end of construction, but they are geared toward providing clients with decision support at key points during a project's lifecycle. Construction industries globally start to rethink what should lie within the scope of services delivered to clients. In most cases there is a clear understanding that structured information that cross-references BIM data to FM provides extensive value-add to them. The questions that still need to be addressed are: Who is appropriating BIM information for use by clients? How does the fee structure change in order to compensate those who do the work accordingly? How can risk be mitigated arising from this redistribution of labor?

The concept behind EIRs is a recent addition to the international BIM jargon. They were first introduced by the UK BIM taskforce in their PAS 1192–2 document. The UK BIM Task Group explains EIRs as follows[1]:

> The EIRs define which models need to be produced at each project stage—together with the required level of detail and definition. These models are key deliverables in the "data drops"—contributing to effective decision making at key stages of the project.

According to the UK BIM taskforce, the three main areas covered in an EIR relate to technical details such as the use of software applications and Model Level of Development (LoD) definitions, details related to management processes, and commercial details related to the content and timing of information handover.

It may appear unusual to ask BIM Managers who are predominantly concerned with the design and delivery of projects to consider those requirements. Truth be said, the EIRs should ideally be declared from the client side. They are the party that's best suited to identify what information would best support their in-house processes for ongoing operation and maintenance of their assets. The conundrum remains that most employers don't yet understand enough about BIM to be in the position to translate their information requirements into a succinct BIM brief. As acknowledged by the UK BIM task group,[2] clients are likely to depend on support from Design Team Leaders and Project Team Leaders to compile their EIRs. They are currently going through a learning process that will enable them to become proficient in developing their EIRs over time. BRE's Paul Oakley puts it this way: "Most clients don't (yet) produce decent EIRs; it is absolutely crucial to make clients aware of the benefits of expressing what they are after. In addition they need to appoint an Information Manager who communicates EIR deliverables and ensures the project team delivers on them!" Ideally, clients introduce the role of Information Manager to the project team in order to manage the flow of BIM-related information they want to get out of BIM in accordance with their declared EIRs. They map those out throughout distinct points in time of the project planning and delivery process. Having someone representing the client in this role adds more clarity to the entire team about the exact BIM output they are working toward.

Understanding a client's EIRs is a crucial starting point, but BIM Managers need to watch out: They can easily get caught out if they confuse what is technically possible with what is commercially wise. BIM Managers need to involve their firm's decision makers in order to determine proper alignment of expected BIM outputs with contractual agreements. If activities leading to fulfilling EIRs fall outside of traditional deliverables, those addressing EIRs should engage in a solid conversation with all affected parties about adjusting the fee accordingly.

There is a reason EIRs are mentioned first in this list: In the past, BIM Managers often overlooked client requirements in the development of BIM support material. Doing so will most likely result in missed opportunities. Working backwards from known client BIM deliverables and adjusting those over time has advantages for a number of other core missions by BIM Managers. For example, clearly expressed EIRs can help BIM Managers to select which parameters to include in the setup of their BIM Library components. Even further, according to the UK PAS 1192–2, EIRs form the basis for suppliers to develop their precontract BIM Execution Plans. EIRs can also provide a useful point of reference for parameter-naming and data-export conventions listed in a firm's BIM Standards. Such standards represent the most obvious terms of reference for any organization that works in BIM.

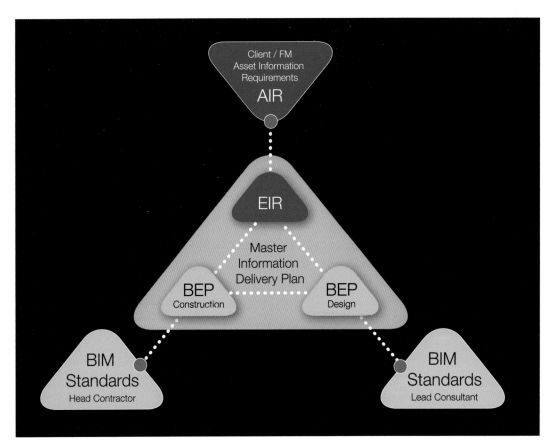

Figure 4–5 EIR in the context of other sources of information.
© Dominik Holzer/AEC Connect

KEY EIR TIPS

1. Don't assume your clients know how to define their EIRs from BIM.

2. Assist them in defining what they are after by showing them how you capture data as part of the BIM process.

3. Break down what information you hand over according to key points of decision making on their side.

4. Expand the dialogue from the client's planning and construction team to their FM service team.

5. Discuss the client's desired pathway for maintaining currency of building data throughout the lifecycle.

6. Establish with the client an asset hierarchy and asset criticality related to key components of your design.

7. List these components in a matrix where you juxtapose them with key properties the client wants to know about.

8. Separate the components by trade and LoD.

9. Adhere to a COBie structure to align the matrix to COBie schemers if required.

10. Use information inherent to the client's EIRs in the way you set up your BIM Execution Plan.

Setting the (BIM) Standards

One of the central actions BIM Managers execute within their organization is to either create or review in-house BIM Standards. These standards are the key reference for anyone who uses BIM within the organization. Standards are more than simple guidelines. In order for standards to make sense, BIM Managers need to declare them unmistakably. They need to be adhered to by all working in BIM, and BIM Managers must continuously monitor and control BIM Standards acceptance. In certain intervals BIM Managers need to revise and update the standards.

> "At the outset of a project we usually have a BIM kick-off meeting where we explain the critical standards to be adhered to by everyone (no matter if they are new to the office or if they have gone through the process already)."
>
> *Alexandra Pollock, Director of Design Technology at FXFOWLE*

BIM Standards govern a range of different rules related to the way BIM gets applied within an organization. Ideally, each individual working in BIM should apply the standards as a matter of course in their day-to-day workflow. Whereas in-house BIM Standards traditionally focused predominantly on bottom-up processes and protocols within an organization, there exists a tendency to reconcile the setup of those standards with the typical EIRs of clients. BIM Standards are usually fairly static and non-project-specific, whereas EIRs are flexible and client-dependent.

BIM Standards cover a collection of different activities within a firm that require a common and uniform approach by all involved in using BIM. These activities relate—among other topics—to project setup, information storing, information naming, and information exchange protocols, as well as the format of 2D (CAD) document output and other forms of deliverables. Standards also regulate the way BIM objects are set up, how they are named, and how they are categorized within the in-house BIM Library.

How to Start

For BIM Managers, standards development may at first be an overwhelming challenge, as many different aspects of BIM authoring need to be considered. So where does one begin? It may work best to start small and to let standards "grow" over time. Luckily for BIM Managers, there is no need to start entirely from scratch when establishing their firm's BIM Standards for the first time. A number of guiding principles and national standards are freely available for consultation. Some fundamental project management aspects in BIM can still be based on previous national or international standards for project delivery in CAD (such as the UK BS 1192–2007). In terms of actual BIM Standards, existing templates appear to be more tailored toward specific software use. The AEC (UK) BIM Standards website[3] offers protocols for a number of different applications. In the United States, the AIA has established three documents in support of digital practice: the E203™—Building Information Modeling and Digital Data Exhibit; the G201™–2013, Project Digital Data Protocol Form; and the G202™–2013, Project Building Information Modeling Protocol Form. Most of the information inherent therein[4] assists in regulating collaborative practice. They are a good point of reference for those who want to align their in-house BIM setup with collaboration protocols. For those using Autodesk's Revit™, the globally disseminated Australia New Zealand Revit Standards (ANZRS) have become a key resource upon which to build. Due to geographical and market differences, BIM Managers should review the content of these guidance documents carefully. They should then establish to which extent information therein can be incorporated directly in their in-house BIM Standards document, or in how far certain passages require prior revision and adjustment.

In those instances where a firm already has a solid set of BIM Standards in place, BIM Managers should still review and update those standards on a regular basis. It pays off to ask those delivering BIM on the floor about their experience and recommendations for updates/edits to the standards. How often should one update? BIM Managers need to understand that too frequent changes within standards should be avoided. Otherwise, employees constantly have to be reskilled with frequent updates. A good interval may be yearly, half-yearly, or whenever there has been a major policy change that needs to be addressed.

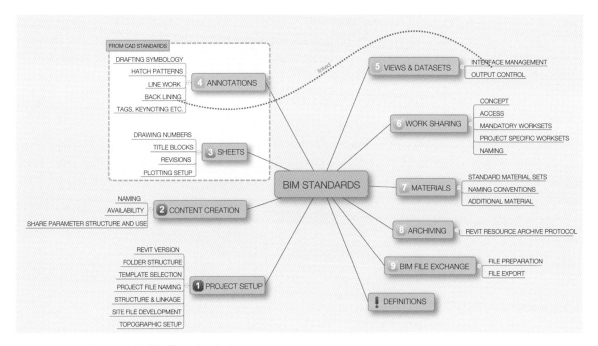

Figure 4–6 BIM Standards diagram.
© Dominik Holzer/AEC Connect

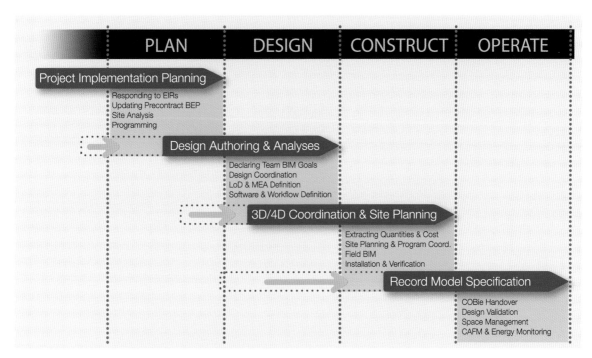

PLAN   DESIGN   CONSTRUCT   OPERATE

Project Implementation Planning
- Responding to EIRs
- Updating Precontract BEP
- Site Analysis
- Programming

Design Authoring & Analyses
- Declaring Team BIM Goals
- Design Coordination
- LoD & MEA Definition
- Software & Workflow Definition

3D/4D Coordination & Site Planning
- Extracting Quantities & Cost
- Site Planning & Program Coord.
- Field BIM
- Installation & Verification

Record Model Specification
- COBie Handover
- Design Validation
- Space Management
- CAFM & Energy Monitoring

Figure 4–8 Typical workflow diagram as outlined in a BEP.
© **Dominik Holzer/AEC Connect**

equivalent Design BIM Execution Plans (DBEP) and Construction BIM Execution Plans (CBEP) are emerging. Another distinction, according to the UK PAS 1192–2, is that BIM Execution Plans get split into precontract and postcontract documents.[7]

What are the key factors needed to go from a good to a great BIM Execution Plan? What are the key sections within those documents that need to be given particular attention in order to maximize their benefits on a project? The development of BIM over the past 10 years points toward a tendency where ever more attention is given to lifecycle aspects of BIM. In the context of BEP, the focus lies on the transition from a predominantly consultants-focused plan to a plan, that also considers coordination by the subcontractors further down the track and, finally, a plan that includes the EIRs for Facilities and Asset Management.

Most collaborators in the BIM space are aware of BEPs and the benefits they provide in regulating the BIM-related workflow. The BEPs have become a pivotal instrument for teams to agree how they would like to approach their BIM journey and what exactly the team is going to deliver. Standard templates for these documents have been around since 2009. Those templates are good starting points from which to work, but the industry is still learning to come to terms with a number of aspects related to the BEP.

One key conundrum associated with the BEP is that one would ideally like to include information that is simply not available at the time when a first issue of the document is needed. One therefore has to account for factors that may strongly affect the collaborative effort, but without actually knowing who will be part of the project team. One needs to second-guess what the requirements of other stakeholders further down the BIM supply chain may be. Consultants don't necessarily engage with the BIM requirements of the subcontractors and neither of these two groups is all too familiar with what the client would like to get out of BIM.

Ideally one would consider a three-stage approach for conceiving BEP templates that can be adjusted for teams that only involve consultants, to teams that span entire consultant/subcontractor/client stakeholder groups. The focus lies in weaving in a level of flexibility and space for evolving over time.

KEY EXECUTION PLAN TIPS

1. Use an existing BEP template and adjust it to fit your project's needs.

2. Align your BEP with information from the EIRs and your in-house BIM Standards.

3. Condense information in a way that captures stakeholders' interest (don't go above 30-40 pages).

4. Ensure buy-in from other key stakeholders and search for their input in generating the BEP.

5. Split between a Design and a Construction BIM Execution Plan (DBEP/CBEP).

6. In defining collaboration, work your way backwards from client requirements—involve the client's Facilities team.

7. Realize that a BEP is a live document that needs to be updated as the project progresses.

The BIM Placemat

A useful tool for BIM Managers to convey the most essential information BIM users within their organization need to know when working a project in BIM is the BIM Placemat. BIM Standards and Execution Plans all play their role in assisting individuals in working toward common goals when using BIM. The problem is that the knowledge embedded in these documents is vast. Publishing declared Standards and Execution Plans are crucial for staff to reference specific clauses or processes. At the same time, colleagues often find themselves in the position where they need the most basic information about running jobs in BIM at their fingertips. In those instances, the BIM Placemat acts like a super-condensed version of the Standards/BEP, all gathered on a single double-sided sheet (placemat). BIM Placemats have a limited information content that fits on a double-sided A3 sheet. The sheet usually gets laminated and finds its way as a core reference at the desk of each employee using BIM.

Putting a single-sheet document together sounds simple enough, but it isn't. Key to the success of establishing a great Placemat is to ask those running projects in BIM about the most relevant bullet-point data they require when starting a project. The responses will most likely appear trivial to any BIM Manager who deals with these issues as a matter of course. Yet, for the common operator, this basic guidance is not always that obvious and a little reminder is immensely helpful.

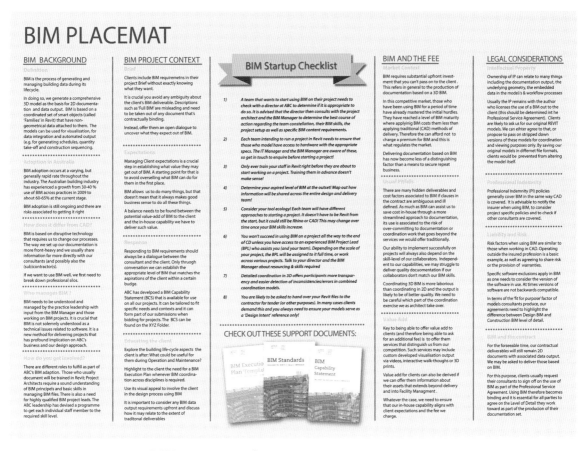

Figure 4–9 BIM Placemat example.
© **Dominik Holzer/AEC Connect**

As an example, the BIM Placemat could remind others about the key reasons behind using BIM. It could then point out how to respond to BIM clauses in a project brief, as well as steps to start a project using BIM (such as notifying upper management and the BIM Manager). Following these topics, the BIM Placemat could provide feedback about staffing and team constellation, setting up a project in the right folders, and so forth. A section of the Placemat could be dedicated to third-party collaboration, another one could highlight typical BIM workflow issues, and another one could tackle specific advice related to the desired format of BIM output. It is best to include a checklist of the 10 most useful steps for implementing BIM on a project. Ultimately, the BIM Placemat does neither substitute the BIM Standards, nor the BEP, but it puts useful information about BIM right at everyone's fingertips.

Gustav Fagerstroem, Senior Technical Designer, explains how at Buro Happold such a document was established: "We worked hard to get what is a very useful document—a one-page document that gives you the basic info. When a new project comes in, you share that crucial information with the entire project team; it doesn't take anyone too long to read it and it lays down the 12 most basic things like: what units do you work in, what tolerances do you work toward, where is project north, who owns what—copy monitoring. These are quite trivial things, but if you get them right from the start, it saves you massive effort later."

KEY PLACEMAT TIPS

1. Start the BIM Placemat by asking your colleagues what information they usually need at their finger-tips.

2. Be precise and to the point.

3. Refer to supplementary documents where necessary.

4. Include checklists for easy consumption.

5. Print out your Placemats, laminate them, and ensure they are on everyone's desks (in addition to being available online).

The BIM Capability Statement

You know how far you have come with your organization's BIM effort, but how do you let outside parties know about it? The capability statement is there to make this explicit and to demonstrate in the most effective way how your firm's BIM infrastructure, staff skill levels, and BIM technology are set up. Being proficient in BIM has become a requirement (in some cases contractual) and demonstrating firm-wide capabilities is an essential part for BIM Managers to promote the in-house setup. The BIM Capability Statement assists project leaders in responding to BIM-specific clauses in the project brief and it helps office managers to cover the BIM side of things when tendering for new jobs.

What Goes in the BIM Capability Statement

BIM Managers should not take lightly the task of producing a Capability Statement. It is important to neither overstate nor undersell your firm's ability to deliver projects in BIM. Clients want to be confident that the team they choose can deliver on BIM. They may not be able to judge the exact extent of a self-assessed promotional document (there are ways to get external bodies such as the BRE in the United Kingdom, or "BIM Excellence" in Australia to undertake third-party assessment). Still, they will likely be able to apply a sound preliminary judgment based on a number of factors to be listed in the Capability Statement.

The most relevant opportunity to evoke the client's confidence is to be able to refer to a number of exemplar projects that have previously been delivered successfully in-house using BIM. Any organization promoting their BIM successes can do so most effectively by pointing to existing projects that were done using BIM. Other

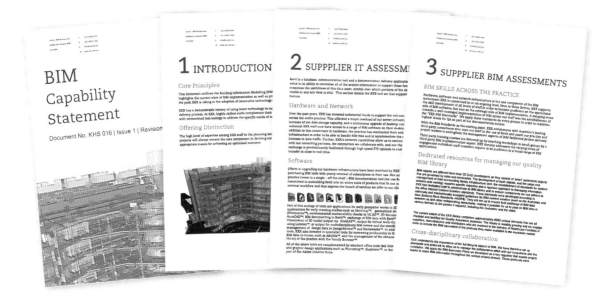

Figure 4–10 BIM Capability Statement content.
© Dominik Holzer/AEC Connect

factors to focus on are staff skill levels and their exposure to BIM in the past. Clients will appreciate being informed about key project team members who demonstrate leadership in the field of BIM.

In addition, the Capability Statement should list in detail the in-house IT/DT setup (servers/BIM authoring seats/connectivity and tool ecologies, BIM Standards, BIM Execution Plans, etc.).

Clients are getting more and more interested in understanding the team dynamic that allows consultants and contractors to work synergistically to deliver projects on time or even early, on or below budget, and without much trouble onsite. For that reason, clients are likely to be interested to know about the overall philosophy that ties an organization's core activity to its use of BIM. Further, clients become more aware of the data-handover opportunities and they may want to see how an organization responds to their needs when it comes to generating BIM output. In that context it is beneficial for an organization to illustrate the range of media used to communicate design and delivery of projects. In some cases this can be accomplished by highlighting the 3D output; in others organizations may add videos, virtual walkthroughs, or Oculus Rift–type interfaces to highlight their output capability from BIM.

Whatever the output format or media, BIM Managers should think about ways to highlight the distinctiveness of the BIM approach their firm offers that puts them in a unique position on the market. One section of the BIM Capability Statement should be dedicated to taking the client step by step through the typical approach to delivery of BIM. The BIM Capability Statement ultimately acts as a firm's "business card" when it comes to BIM. It should be set up in a way that allows it to be used as a template to feed into a number of other documents for promotion to clients or on project tenders.

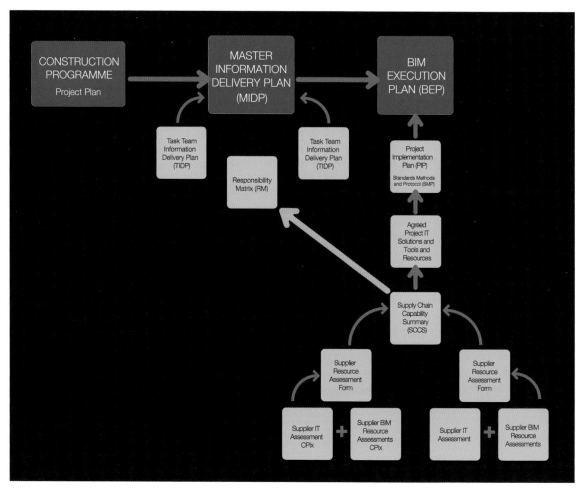

Figure 4–11 BIM workflow as described in the UK PAS 1192–2.
© Dominik Holzer/AEC Connect

UK PAS 1192–2/3/4/5-Specific Documents

The information provided about BIM Capability Statements so far is of a general nature, useful to all. The UK BIM taskforce regards the inclusion of the supply chain capability assessment as a strategic part of the Construction Project Information Exchange (CPIx) protocol and other specifications that form part of their PAS 1192 series of documents.

Key components of the CPIx are the Supplier BIM assessment and the Supplier IT assessment that get consolidated in a Supplier Resource Assessment Form. The BIM Assessment Form asks four gateway questions: (1) the willingness to exchange data and the quality of that data; (2) nominating 12 areas of BIM from which the project will benefit; (3) BIM project experience based on up to three reference projects; and (4) a BIM Capability Questionnaire.

Information contained in the CPIx documents gets passed on directly into the setup of a Project Implementation Plan (PIP). It ultimately feeds into the generation of the pretender BEP. In other words, the BEP becomes the point of convergence between the declared BIM capabilities of those contributing to the BIM effort on a project on one end, and the client's declared BIM requirements on the other.

BIM Library Management

BIM Managers are responsible for setting up the structure, procurement, population, and management of their firm's BIM Library. What looks at first like a straightforward task is often very time consuming and requires discipline and rigor to achieve. A library is a live source of knowledge and BIM Managers are responsible for ensuring that it is up-to-date and that information can be found by those searching in the shortest time necessary via a centralized site. They need to extract new and useful content from all ongoing or recently finished projects; they need to ensure it conforms to the Content Creation Standards as prescribed in the BIM Standards. They then need to ensure that content gets certified and added to the "official" library as a resource for all to access. As part of that, BIM Managers determine the exact categories into which content is to be classified. That way, they maximize usefulness of the library and minimize the time required to find any specific BIM object in the library.

Figure 4–12 BIM Library wordle.
© Dominik Holzer/AEC Connect

Spotlight on BIM Content

For most organizations, the quality of their BIM output is highly dependent on the quality of content within their BIM Library. It is a paradox fact that despite this, a firm's BIM Library often gets neglected, with little or no resources allocated to their establishment and ongoing management. Why is this?

Setting up a well-configured BIM Library comes down to the BIM Manager's ability to communicate the relevance of a quality library to the firm's leaders and to those who generate (and require) BIM content. As a starting point, a BIM Manager (or a dedicated BIM librarian working alongside the BIM Manager) should investigate what BIM content is already available, and what content is most fundamentally required within an organization. They then classify items according to priority in order to determine gaps in the library that need to be addressed. Overall, such a strategic approach will help the BIM Manager to identify the most immediate (and often urgent) requests for new content by project teams, while establishing the broader needs of an organization.

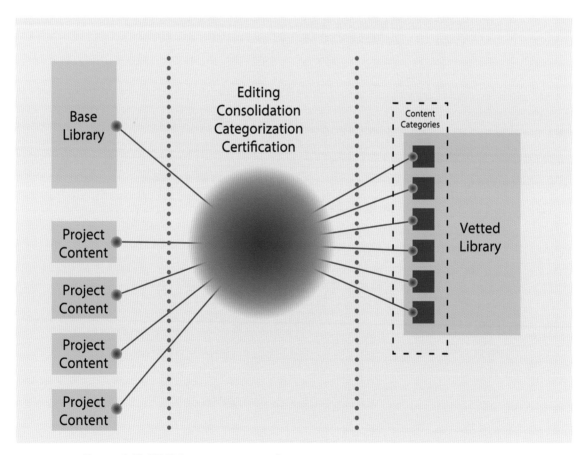

Figure 4–13 BIM Library structure and categories.

© Dominik Holzer/AEC Connect

Yet, a strategic approach to setting up a BIM Library is not always easy to accomplish. In a highly project-focused environment such as the construction industry, knowledge management is often overshadowed by ad hoc project milestones and deadlines. Even though a sophisticated support infrastructure for BIM such as the BIM Library is pivotal for any organization, it is nevertheless not always given the priority it deserves. One reason for that neglect may be that practices put equal emphasis on their BIM Libraries as they did with their CAD libraries in the past. This is a major mistake. Whereas a not-so-perfectly organized CAD library would have possibly slowed down the CAD documentation process a bit, a badly managed BIM Library can severely impede or even delay the documentation process.

The BIM Library: Typical Problems

One potential problem with a BIM Library is that its content can end up all over the place if those contributing to the library do not adhere to a disciplined approach. Sourcing useful library content is one thing; ensuring it is named correctly and sorted against a dedicated object category is another. This conundrum reveals a further issue faced by those who procure BIM Libraries: There are usually multiple users contributing objects to the library. Often these objects stem from BIM authors who work on different projects or they are sourced from third-party BIM object suppliers. This multitude of input options lends itself to multiple entries of the same object under a different name and category. If there is no certification mechanism in place, some library objects may render BIM files unusable as they use an excessive polygon count that makes a BIM file inoperable. Next to the shortcomings listed above, inadequate library management may also result in the use of inadequate parameters associated with BIM components, which in return can result in loss of data fidelity when scheduling room data or extracting 2D information to plot-sheets from the model via View templates.

There is a lot that BIM Managers can and should do to address the above issues. First and foremost, any firm using BIM needs to have a strategy in place that regulates who generates or sources BIM content and how it gets

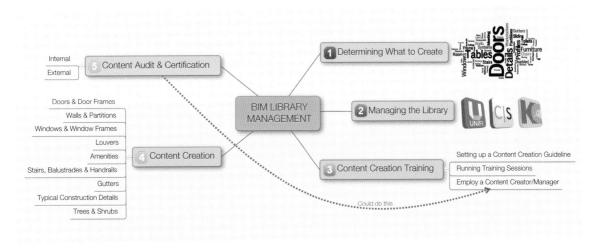

Figure 4–14 BIM Content strategy diagram.
© Dominik Holzer/AEC Connect

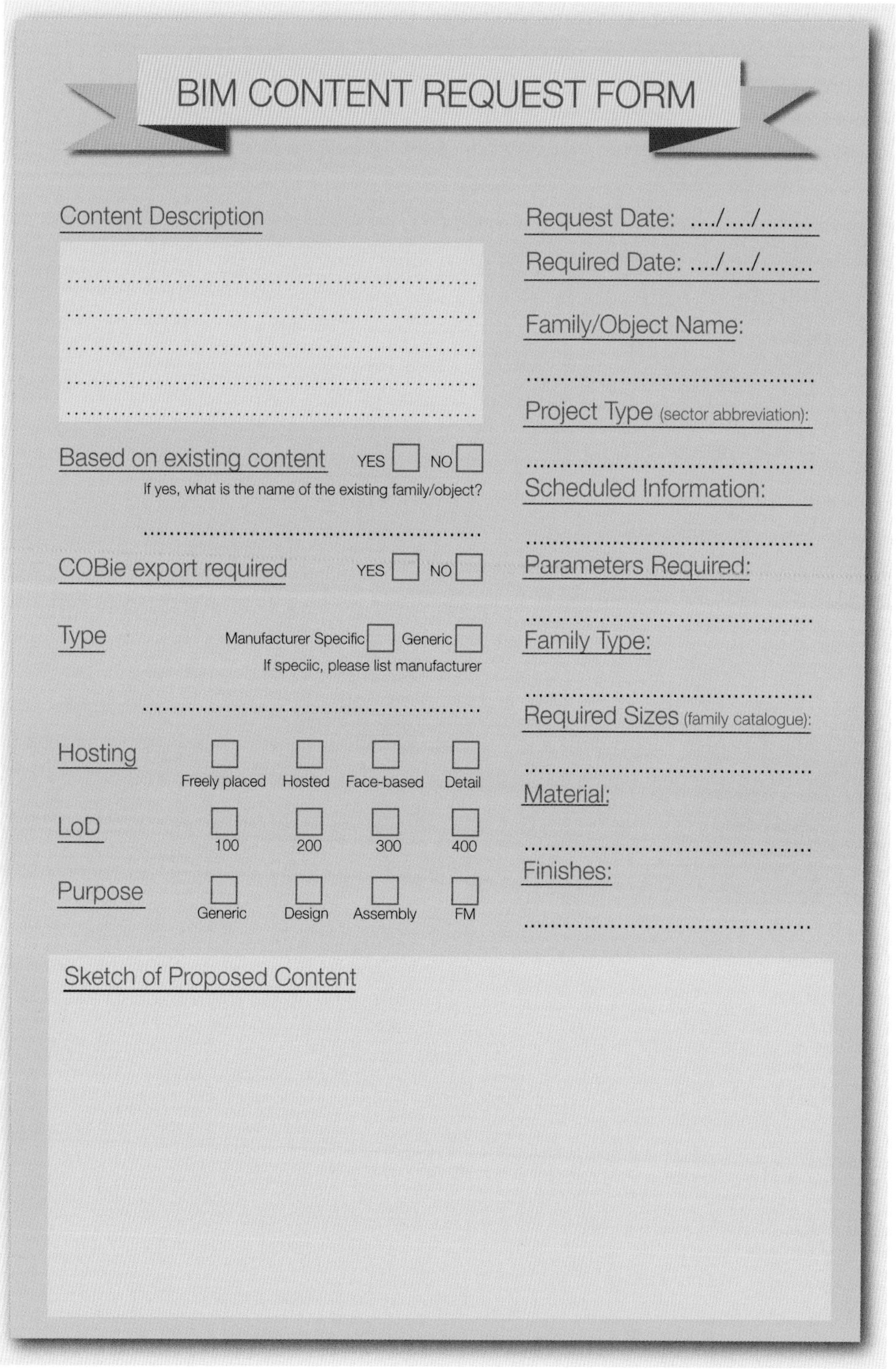

BIM CONTENT REQUEST FORM

Content Description

....
....
....
....
....

Based on existing content   YES ☐   NO ☐
If yes, what is the name of the existing family/object?

....

COBie export required   YES ☐   NO ☐

Type   Manufacturer Specific ☐   Generic ☐
If speciic, please list manufacturer

....

Hosting   ☐   ☐   ☐   ☐
            Freely placed Hosted Face-based Detail

LoD   ☐   ☐   ☐   ☐
          100   200   300   400

Purpose   ☐   ☐   ☐   ☐
            Generic   Design   Assembly   FM

Request Date: ..../..../........
Required Date: ..../..../........

Family/Object Name:

....

Project Type (sector abbreviation):

....

Scheduled Information:

....

Parameters Required:

....

Family Type:

....

Required Sizes (family catalogue):

....

Material:

....

Finishes:

....

Sketch of Proposed Content

Figure 4–15 Example of a BIM Content Request Form.

© **Dominik Holzer/AEC Connect**

added to a centralized library. BIM Managers are not necessarily the party producing content (even though they are likely to contribute to the content-creation effort). They need to supervise and manage the information flow in and out of libraries and across all projects that are delivered using BIM in a firm. One way of achieving high standards in content management is to nominate key individuals who contribute to the BIM Library. BIM Managers then set up a hierarchy that outlines how content created on projects finds its way into a certified library that is accessible to all. In some cases, an organization may be in the position to nominate a dedicated BIM Librarian who works with the BIM Manager to procure the BIM Library across a firm. In larger firms the BIM Manager sets up a structure of how object information stemming from those authoring BIM is collected (e.g., on a weekly basis) by the project's Model Manager to then be certified by them (or a BIM Librarian) and passed on to the centralized library. In smaller firms the BIM Manager him/herself will need to audit existing projects on a regular basis to extract any content that may be useful, to then adjust and certify it for inclusion to a centralized library.

The Content Request Form

BIM Managers should also ensure a system is set up for individuals in a practice to request the creation of bespoke content on a project. Such requests should be submitted to the BIM Manager (or the BIM Librarian) in a standardized format according to a predefined template. That way, they avoid having to second-guess what the requester is after. The key queries to be listed on the request sheet are as follows.

BIM CONTENT REQUEST SHEET INCLUSIONS

- Content description
- Request date & date required
- Based on existing (1) or new (2). If (1), what is the name of the existing family/object?
- Requires creation or support for creation?
- Family/Object name (must follow standards)
- Project type (sector abbreviation)
- Scheduled information required
- Parameters required (codes, dimensions)
- Family type
- Required sizes (family catalogue)
- Intent/Purpose (sketch, generic, design, manufacturer/assembly, asset management)
- Attachments (product information)
- Generic (1) or Manufacturer specific (2). If (2), list manufacturer
- Level of Development (LoD)
- Visibility

- Material

- Finishes

- Specification

- Hosting (Freely placed/Hosted/Face-based (light fixture)/Detail component maps to geometry masks)

- MEP connectors

- COBie export required

The request sheet should also contain an area to allow the party requesting content to highlight the following information:

- Origin and flip controls definition

- Additional subcategories? (To help control the visibility of family geometry)

- Parametric behavior

One way to verify if a BIM Content Library is set up properly is to check the time it takes for BIM Authors to find the content they are after. Those working on tight timelines can't afford to comb through hundreds or thousands of objects by browsing through messy folders hoping to get lucky. Instead they want to have the content they need at their fingertips by using a limited set of keywords that allows them to find what they are after. Associating Families/Content to a predefined set of categories with the correct nomenclature is a key step in facilitating an efficient search process. Adding thumbnail views of the content (that one can skim through via a browser) speeds up the search process.

Preparing for Content Management

In those instances where a BIM Manager starts setting up a BIM Content Library, or where a library undergoes a major revision, a scoping exercise should precede the library's setup. The BIM Manager then produces a Content Management Guideline to communicate to the key goals of content creation, management, and storage in the BIM Library to the practice.

A BIM Content Creation Guideline can contain the following key points:

1. The firm's BIM Standards for BIM Content creation and certification

2. The management of the BIM Content Library in terms of assistance with the technical setup of the central BIM Library, the audit of its existing Families/Objects, methods for upgrading audited elements to the most recent BIM authoring version, updating staff about newly available content, and an investigation into BIM Library Management tools (as described in Chapter 3).

3. Adding new content by creating or sourcing content based on the gaps in the central BIM Library that were identified earlier.

4. Providing project content support by determining content requirements during project phases.

5. Ongoing Audit of Project Content. What do BIM Managers or BIM Content Librarians look for during their regular skimming of live projects for newly developed content?

6. The Content Creation Guide includes a step-by-step tutorial of how content should be set up according to the firm's BIM Standards.

7. BIM Managers or Content Librarians should complement the Content Creation Guide by conducting targeted content creation training. In order to maximize usefulness of these sessions the BIM Manager/Content Librarian canvass feedback from staff about the key topics in preparation for the training. By hosting regular training events and by posting a referring FAQ list online, they can further improve the skills of those producing a firm's BIM content.

8. Overall, it is the BIM Manager's/Librarian's task to transition a practice from having staff with a high dependency on third-party content experts, to becoming self-supporting on projects over a period of time.

How to Train for BIM

Let's imagine the following scenario: A firm wants to use BIM on a project and the BIM Manager is asked to ensure that a group of designers who have never been exposed to BIM will cope with their task at hand. How

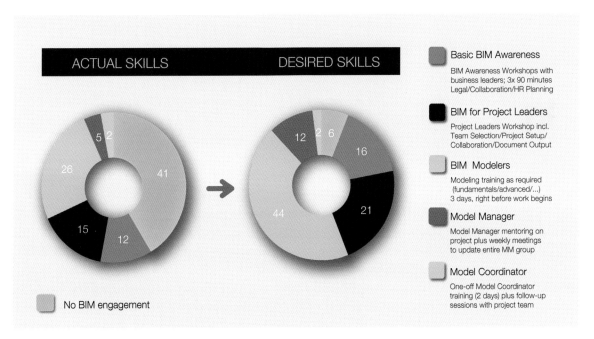

Figure 4–16 BIM training strategy diagram.
© Dominik Holzer/AEC Connect

can BIM Managers succeed in empowering colleagues to master the transition from CAD to working in the unfamiliar context of BIM? What are the options? What skills should they develop first? What role does the BIM Manager play in empowering others and mentoring them on their path of discovery and learning?

BIM skills development is a fundamental task for any BIM Manager. The provision of guidance documents/templates and the setup of a well-structured BIM Library form the backbone of any sound BIM support infrastructure. Setting up a focused BIM training regime complements these support efforts and it is a logical step as part of a BIM Manager's Change Facilitation in an organization. The ultimate goal of any training is to assist those trained to help themselves.

There are a number of misconceptions about how to best train in BIM. In short: There is no simple answer. The right way to train BIM always depends on the context, the type of work that those undergoing training are expected to do, the setup within a firm in terms of dedicated equipment/rooms for training, and much more. It is also important to understand who gets trained and when is the best time to do so.

Get the Timing Right

A strategy for training always requires a BIM Manager to understand the dynamics apparent in the firm that lead to starting projects using BIM. The timing and content of any training should be aligned with specific project needs. A typical mistake BIM Managers make is to pretrain staff (often pushed by upper management to do so) with the expectation that the trainees will be ready to hit the ground once a project eventuates. Pretraining usually occurs when project load is low and a two- to three-day learning module does not signify a major disruption to anyone's schedule. As much as this approach appears reasonable at first, it is a recipe for disaster in most cases. Research in "forgetting behavior"[8] highlights that on average learners forget about 40 percent of what they learn within one day. The rate of forgetting depends on a number of factors, in particular the rate of interventions that aim to improve long-term remembering such as repetitions in certain intervals. In other words, if a future BIM operator undergoes training without strategic follow-ups or direct progression of applying what they learned "in heat," chances are that he/she will have forgotten the majority of what was learned within a week or two. Pretraining without continuous follow-up (e.g., on a hypothetical case study) does not work; it is simply a waste of time. Even more, it reflects badly on those who undertook training in the first place as they will be very badly equipped to be productive on a project once it starts. BIM Managers need to communicate this issue clearly to project leaders and upper management. They need to ensure that BIM newcomers starting on a project are trained right before their involvement. But there is more. They also need to communicate that the decision to have someone starting in BIM should not be short-sighted. Having a BIM newcomer working on a BIM project for two to three months only for that person then to revert back into working in CAD for a prolonged period will result in problems further down the track. First, the operator will lose some BIM skills over time. Second, the person concerned will likely get frustrated "being degraded" to working in CAD once exposed to the benefits of working in BIM. Third, by jumping between BIM and traditional work methods, the practice will lose out on an opportunity to build up BIM capacity across the organization.

Programming BIM Training

It is a sad fact that many BIM Managers still confuse BIM training with BIM software training. As much as software forms an important factor for those who eventually author and coordinate BIMs, software skills are but one aspect within a range of other things to know about BIM. In order to program BIM training appropriately, BIM Managers need to step away from a too narrow focus on software training courses and focus on the skill development of individual staff members instead. In that sense, they need to reverse the questions and ask: What does this person in his/her particular role need to know about BIM in order to do their job well? That way the development of BIM capacity across an organization starts with a personal journey of skill acquisition. What is their desired trajectory to use BIM proficiently in their job and how do they get there? Such considerations form part of a BIM Manager's Change Management strategy. In practical terms it means that the BIM Manager needs to assemble different training modules (to be delivered either in-house or externally) that have distinct focus on various aspects related to BIM delivery. Some should focus entirely on what Practice Leaders should know about BIM (e.g., contractual and workflow issues, fee-related issues, etc.). Other modules may focus on BIM for Project Leaders (how to respond to BIM clauses in briefs, staffing for BIM team selection, BIM process explained). Then there are distinctive training modules depending on the type of modeling required (e.g., BIM for town planning/massing studies, architectural design, interior design, cost extraction, field BIM, . . .). Next, the BIM Manager needs to consider distinct training modules for beginners and advanced users. These modules then need to be matched against individual colleagues and their distinctive learning paths. In some cases, different modules will need to be delivered by different instructors either in-house or externally.

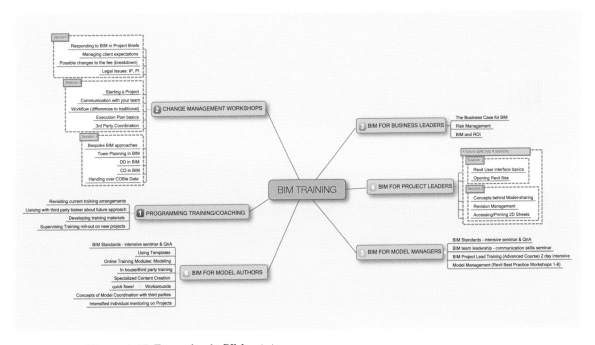

Figure 4–17 Example of a BIM training program.
© Dominik Holzer/AEC Connect

Training: In-House, External, or Online

There exist many different views about how to best organize the delivery of BIM training. Traditionally, the question has revolved mostly about a distinction between in-house learning and outsourcing it to third-party providers. More recently e-learning via online courses has been added to the mix and BIM Managers are tasked to find the right approach for their specific organizations. This may well involve a combination of two or even all three training delivery methods.

Each of the options has its own pros and cons. It is important for BIM Managers to understand these to then make an informed judgment on how to proceed.

Depending on the size of a firm, in-house training is the obvious choice. For smaller and medium-sized practices, it is easier to put time aside to run training sessions within their offices. Issues to consider here relate to the fact that in-house training requires the right infrastructure—such as a dedicated training space—to be available. Experience shows that in such contexts the danger lurks that staff get easily sidetracked by their day-to-day duties during training. It is easy to lose focus if those attending the training are not sufficiently disciplined to stick out the two to three days usually recommended to go through a focused training module. Depending on what needs to be learned, training sessions can also be spread over a longer period with shorter sessions. This is not advisable; when it comes to learning how to model or to coordinate BIM (which constitutes the most common form of BIM training), a consolidated effort across two to three days makes the most sense. The danger remains that training is not delivered in a focused fashion, but rather as a series of "firefighting" exercises where those to be trained oscillate between project deadlines while picking up some tricks on the side.

The advantage of in-house training is that an organization's actual project template can be used during the training regime. Operators get trained in the working environment they will later be using as part of their project work. The firm's own user interface, organization-specific templates, and other settings can all be used, which, in return, will provide a more familiar context to those who go from training directly to working on a project.

Having staff trained externally has its own set of advantages and disadvantages. On the one hand, it frees up the time of the BIM Manager who can rely on staff getting educated by experienced training providers in a dedicated environment. Many larger organizations choose to get their staff trained that way. Depending on the amount of staff to be trained per session and the relation to the third-party training provider, firms can request for their own templates to be used (thereby diminishing the gap between "general" BIM training and a more bespoke focus on how BIM gets applied within their organization). The cost factor of external training may prevent some from choosing this option, as well as the dependency on availability of external training providers (and training spaces). BIM Managers sessions are often hard to schedule due to the short notice they usually receive. There is little a BIM Manager can do to predict training needs due to the uncertainties related to staffing on projects where BIM skills may be a last-minute requirement. As stated earlier, the pretraining option is not advisable.

Emerging additions/alternatives to in-house or external training are online training courses. These courses have gained in popularity, in particular for those aspects of BIM that focus predominantly on learning the features inherent to BIM authoring and coordination software. Online courses make a lot of sense if they can be effectively monitored so that staff actually complete them in a timely fashion. It is the task of the BIM Manager to fill the gaps between knowledge gained by trainees online and the use of BIM "in-heat" on projects. If BIM Managers choose to rely heavily on online tutorials, they need to appropriate the available material (course selection/configuration) and follow up with staff, providing complementary tips and tricks right after they go through the tutorials. In addition, it will help those who went through online training if the BIM Manager acts as a mentor during project startup and at key intervals during the project. An alternative to general online training is for BIM Managers to record and dub their own training tutorials via screen capture tools such as the free CamStudio™ application. This process can be a time-consuming exercise for BIM Managers, in particular during the initial trials. It, nevertheless, offers BIM Managers an additional option to share their knowledge in a highly targeted fashion without having to repeat the same information over and over again.

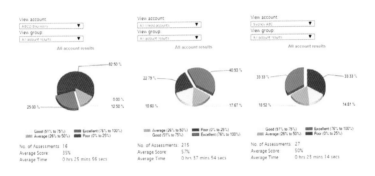

Figure 4–18a KnowledgeSmart Comparison Chart.
© KnowledgeSmart

Figure 4–18b KnowledgeSmart User Page.
© KnowledgeSmart

Skill Assessment and Ongoing Development

Some organizations are keen to investigate how their staff progress in their engagement with BIM. In relation to understanding BIM as a process, as well as understanding the quality of the setup of their BIM infrastructure, BIM audits, as described in Chapter 2, are the most effective instruments to accommodate this. If a practice wants to test the BIM modeling capability of its staff, it can check these skills via third-party assessment such as Knowledge-eSmart. These tests are usually done either when recruiting new staff or during regular intervals to check the skills of existing staff. Any feedback provided via those assessment institutions can help the BIM Manager to direct his/her focus on particular areas in BIM skill development across an organization. Based on the responses received via the assessment approach mentioned here, BIM Managers can create a "Company Skills Matrix" to map skills against tasks across an organization. Gaps can thereby be identified that will allow the BIM Manager to communicate any business case for further investment into staff skill capability or the recruiting of new staff with the desired skill sets.

Reaching Out

"My job is to work myself out of a job!"

Brok Howard, Firmwide Design Technology Specialist at HOK

This radical assessment sums up a fundamental principle behind the support provided by BIM Managers: A large portion of a BIM Manager's work relies on helping others to help themselves. BIM Managers need to reach out internally and communicate progress and challenges related to the firm's BIM efforts. As stated before, putting a number of text documents on the server for others to download doesn't cut it. BIM Managers need to nourish a culture of active engagement with BIM across their practice.

BIM User Meetings, Newsletters, and DT Websites

The best way to get people on the floor involved is to hold regular BIM user meetings either at lunchtime or during work hours (depending on office policy). Another instrument to keep staff up to date with the latest developments on the DT/BIM side is a regular BIM newsletter. There, the BIM Managers briefly mention the latest achievements, updates to the library, interesting articles about DT/BIM, and possibly some tips and tricks for staff to consider. The newsletter should not overload others with material; it should be using engaging images, with as little text as necessary. Staff should be able to click on thumbnails in the newsletter to then be taken to an intranet website where more detailed information, videos, documents, and so forth are displayed.

Spreading the BIM Love

"We have been successful to 'flip a studio' by having an internal expert joining them for a period of time to facilitate change."

Shane Burger, Global Design Technology Director at Woods Bagot

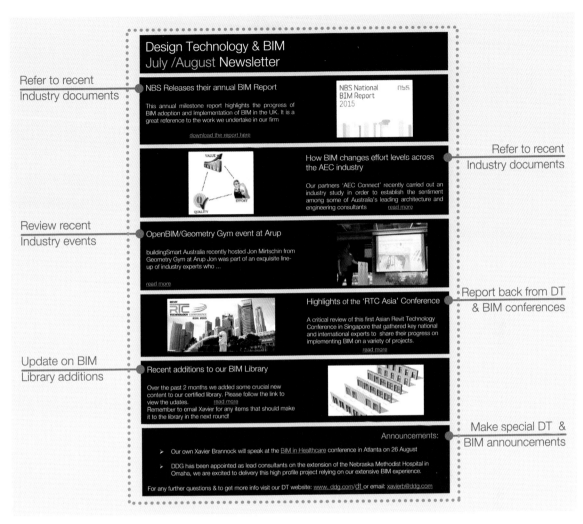

Figure 4-19 Example of a BIM newsletter.
© Dominik Holzer/AEC Connect

Depending on a firm's size, constellation, and geographic distribution, one single BIM Manager can't get the word out alone. It is common that BIM Managers rely on (and at times strategically place) local BIM representatives, working with project BIM leaders (or Model Managers) who look after BIM on a single, or across a number of projects. They become the go-to people when BIM Authors or coordinators hit a problem. This middle layer of BIM support should get together with the project team on a weekly basis to discuss any issues experienced. Together with the BIM Manager, they form a firm's "BIM Team" that shares responsibilities for rolling out BIM across an organization. As with any form of collaboration, it is crucial for a firm to set up a reporting structure that regulates the information flow between BIM Team members. The associated reporting structure does not necessarily need to conform to a hierarchical/geographical format; it can be split up into more activity-focused sections for various BIM Managers to contribute their knowledge to.

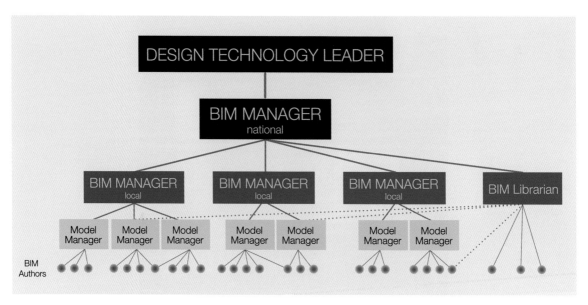

Figure 4–20 Spreading workload across the BIM and Design Technology team.
© Dominik Holzer/AEC Connect

More about the daily duties of the BIM Team and typical pathways for interaction will be discussed in Chapter 5. The "back of house" material presented here constitutes the backbone for these daily operations to unfold seamlessly. The next chapter is entirely dedicated to pointing out how BIM Managers deal with their day-to-day affordances.

Thank you to all the experts who so generously offered their thoughts and insights for this chapter: Lee Wyles of BDP, Paul Oakley of BRE, Alexandra Pollock of FXFOWLE, Jan Leenknegt of BIG, Gustav Fagerstroem of Buro Happold, Shane Burger of Woods Bagot, and Brok Howard of HOK.

Endnotes

1. http://www.bimtaskgroup.org/bim-eirs/, accessed March 22, 2015.
2. http://www.bimtaskgroup.org/bim-eir-faqs/, accessed March 22, 2015.
3. https://aecuk.wordpress.com/documents/, accessed April 17, 2015.
4. http://www.aia.org/groups/aia/documents/pdf/aiab095711.pdf, accessed April 2, 2015.
5. https://bimforum.org/lod/, accessed April 12, 2015.
6. https://www.consensusdocs.org/News/ViewArticle?article=Podcast-on-ConsensusDocs-301-BIM-Addendum, accessed March 14, 2015.
7. http://www.designingbuildings.co.uk/wiki/BIM_execution_plan_BEP, accessed March 22, 2015.
8. Loftus, Geoffrey R. (1985). "Evaluating Forgetting Curves." *Journal of Experimental Psychology: Learning, Memory, and Cognition*, 11. 2:397–406.

DAY-TO-DAY BIM MANAGE-MENT

BIM Management is a highly interactive process with a great variety of different tasks to be accomplished on a daily basis. This chapter reviews these day-to-day activities and explains how BIM Managers master them most efficiently. It looks at in-house requirements, as well as the necessity for integration and coordination of BIM data across a multidisciplinary project team. It highlights how interpersonal and communication skills, as well as the ability to formulate concise business plans, are fundamental to a BIM Manager's role. Whether they get applied during mentoring of other staff or during large project coordination meetings, the BIM Manager needs to be articulate in expressing his or her expert view clearly and effectively. In the long term, moreover, it emphasizes how BIM Managers need to establish a culture of dialogue and, to a degree, peer-to-peer support with the goal of disseminating BIM knowledge across their entire organization (and beyond).

Figure 5–1 University of Nottingham Technology Entrepreneurship Centre (TEC), Nottingham, UK.

Copyright © Bond Bryan Architects LTD

Most BIM Managers will empathize with the following scenario: Arriving first thing in the morning at their desk for the day, they put together a "to do" list to tackle, but their plan of action is put into immediate jeopardy by unknown and unanticipated events and challenges. It is as if the BIM Manager's role and ability to execute his or her workload efficiently is always characterized by an undercurrent of unease and uncertainty.

The common state of flux highlights a fundamental issue for BIM Managers: BIM Management in daily practice very often combines a portion of strategic development and advancement with reactive support and fire-fighting. One might argue that this is simply a matter of fact and BIM Managers ought to accept it and deal with it as best they can. There may be some truth to that, but at the same time there are mechanisms and strategies BIM Managers can apply to diminish how much of their time is spent on reactive and ad-hoc support. This kind of assistance is not problematic as such, but it can become highly disruptive to other, well-managed undertakings; it also takes over large portions of a BIM Manager's time. The solution to this problem often comes down to a clear articulation and distribution of responsibilities and associated time management. It also comes down to being able to rely on a great team to assist with the daily workload. Subdividing BIM-related tasks is easier said than done: The range of deliverables BIM Managers are trusted with is usually ill-defined. With a lack of definition comes a lack of opportunity to manage these tasks, delegate parts of them to others, and get a grip of the time required to fulfill them.

BIM Management doesn't have to be an uphill battle; the key to success is proper planning and the alignment of BIM-related management activities with the overall business strategy of any organization. For that to happen, BIM Managers need to learn how to delegate, how to develop concise business plans, and "sell" those to their leadership for buy-in. As stated in Chapter 2, "Change Management," those overseeing the implementation of BIM are often not equipped with skills that would justify calling them "managers." More likely they would have grown into this role one way or another (either via a vested interest in technology, or simply because they were able to cope well with BIM-related software). A first step to coping with day-by-day BIM Management therefore is to acknowledge that a structured approach to managing time, resources, and workload is essential.

The Broad Spectrum of BIM

There exists no "checklist for BIM Management," and this is for a good reason: By now there are simply too many different activities associated with BIM Management across an entire spectrum of stakeholders. A single list would not apply to any given situation as different BIM Managers from different organizations and stakeholders throughout the supply chain have different priorities to deal with. What BIM Management means for any one individual must therefore refer to the fulfillment of their organization's core business, plus the facilitation of collaborative goals on a project.

Within their individual role, BIM Managers can clock up dozens of different tasks to fulfill. Many of those will be required on a day-to-day basis. It is still rather uncommon for BIM Managers to be presented with a concise role description at the outset of their employment (although some advanced BIM users have now established distinctive roles within their BIM Management team). With the ever-changing context of technology associated with BIM, roles and responsibilities are changing and upper management simply doesn't know what exactly to request from their BIM team. Most practices rely on their BIM Managers to work out and define their own area of responsibilities and the degree to which they support (and get supported by) the design or construction team, the leadership team, and the IT department. With the lack of a clear boundary about what their role entails, BIM Managers often struggle to understand where to start. The following list provides a structure to work toward by distinguishing between four major groups of tasks:

1. Strategic tasks for the advancement of BIM across the organization

2. Project-specific tasks that can be anticipated

3. Project-specific tasks that cannot be anticipated

4. Activities associated to their own learning and skill development

Day-to-day strategic BIM deliverables encompass setting up a BIM support team, whereby BIM Managers establish the line and frequency of reporting with others who collaborate closely on procuring BIM within a firm. Hand in hand with these tasks goes the definition of business cases and the communication to upper management. Strategic undertakings also involve Change Management (see Chapter 2) development of back-of-

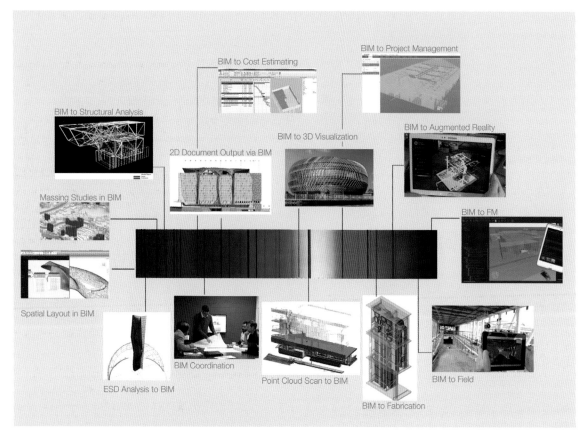

Figure 5–2 Potential BIM spectrum.
Copyright © Dominik Holzer/AEC Connect

house support such as BIM Standards and templates development (as detailed in Chapter 4, "Building Up a BIM Support Infrastructure"), managing the BIM content library (including the search for and purchase of third-party content), programming/carrying out training, establishing an annual BIM and Design Technology budget, testing new tool releases, and interfacing with IT.

BIM planning on projects is the main topic covered in this chapter; it includes a range of tasks such as:

- Reviewing project BIM requirements in the project brief

- Assisting the setup of BIM-enabled project teams

- Partaking in regular meetings for resource planning

- Developing tool ecologies that target the desired project output (described in Chapter 3, "Focus on Technology")

- Adjusting BIM Execution Plans to suit a specific project context

- Setting up the modeling context for teams, providing startup mentoring on projects

- Partaking in regular project reviews with the Model Manager (if applicable) and those authoring models

- Overseeing project-specific content creation and certification
- Assisting project teams with specific submissions, partaking in BIM coordination sessions (if applicable)
- Assisting with the proliferation of model information into complementary output for 3D visualization, quantity takeoff, costing, and so forth
- Site programming
- Quality Assurance (QA) of BIM Models and their resulting document output
- Linking BIM to FM and more

The reactive resolution of unexpected issues usually causes the most headaches for BIM Managers. The time required to fulfill tasks associated with it fluctuates and is therefore hard to capture. In short, project-specific support that cannot be anticipated equals fire-fighting.

The fourth group of tasks, namely the advancement of their own skills and the promotion of BIM excellence to the outside world, is covered in Chapter 6, "Excelling Your BIM Efforts."

Advancing BIM Strategically

The strategic component of daily BIM work can most easily be defined as it relates to typically well-known activities. Many of these are described in other chapters as they relate to the cultural aspects of BIM and Change Management, as well as tool selection and back-of-house activities. This section will position the BIM Manager as operational information integrator on the floor. In order to do so, it is first important to situate the BIM Manager among his or her colleagues as this will assist in comprehending not only responsibilities, but also lines of communication.

Setting Up the BIM Team

The strategic advancement of BIM is closely tied to an organization's structure. In mid-sized to large organizations BIM Management usually is a team effort and the ability to delegate and distribute responsibilities and accountabilities across the team is a crucial component of the BIM Manager's undertakings.

With the increasing proliferation of BIM in all sectors of design, engineering, construction, and operation, daily BIM tasks typically get added to the workload undertaken by project architects and engineers, as well as design/construction managers. For now, the BIM Manager more commonly represents a layer of support for teams, or possibly may combine the role of support with more design/engineering or design/construction management activities. In general, the BIM Manager is referred to as the person who is mainly responsible for procuring and advancing BIM use across an organization. He or she doesn't do so in isolation, but most likely reports to a (Design) Technology leader. In larger organizations with multiple (often remote) locations, a BIM Manager is likely to interact with several others and Model Managers/BIM Project Leads in running the BIM team. Any medium- to large-scale organization is well advised to set up a structure for their BIM team to

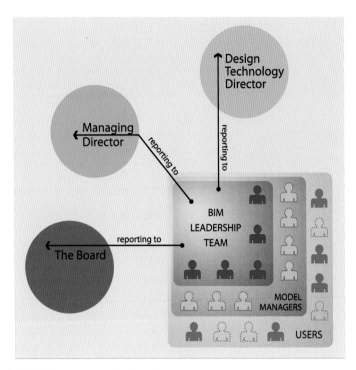

Figure 5-3 BIM team organizational structure.
Copyright © Dominik Holzer/AEC Connect

regulate typical lines of communication: Reporting of BIM-related issues from the floor and communicating key strategic aspects of BIM to upper management. The BIM Manager is likely to be the information hub that empowers both sides to cope with the impact of BIM on their primary tasks, be it running projects or running the business (and anything in between).

BIM Managers Facilitating Communication

In their role as communicators, it falls under the BIM Manager's duties to organize regular user-group meetings, and educational seminars. In order to maximize the benefits of such sessions, BIM Managers can first canvas ideas among staff to then target specific topics that others want to talk about (this could relate to specific features within the software, to updates/changes to the BIM Standards, or presentations of successful projects including lessons learned, etc.).

These sessions help to establish BIM as part of the overall activities of the organization and they generate a greater sense of purposefulness and belonging to those who are involved in the firm's BIM efforts. As much as the BIM Manager should moderate these sessions, there should also be a portion of freedom for the floor to express their views, exchange ideas, and suggest changes to the current BIM strategy. In other words, BIM Managers need to take these sessions as an opportunity to listen to others and understand the mood on the floor. It is crucial for these sessions to remain on the office agenda; with BIM Managers easily side-tracked

(a)

(b)

Figures 5–4a, b   BIM team meetings.
Copyright © Dominik Holzer/AEC Connect

Figure 5–5   BIM coordination meeting.
Copyright © Point Advisory

(or overwhelmed by project-related deadlines) the danger lurks that user-group meetings fall off the radar or simply get suspended for long stretches of time. When this happens, BIM Managers run the risk of losing the grip of what happens on the floor, communication is disrupted, and BIM progress across an organization stalls.

Next to in-house user-group meetings, BIM Managers typically schedule regular meetings with other BIM team members to get an update on current developments throughout the organization. This is particularly relevant where an organization works from several geographically disperse offices. In those instances there is likely to be a "national" manager who coordinates a group of local/state managers. Members from the IT team typically join regular meetings of the BIM team as they often relate to the planning of resources across Information Technology (IT) and Design Technology (DT).

One other area of responsibility of the BIM Manager is the assistance to the firm's HR department in defining job descriptions for new positions that require BIM skills as part of their employment—BIM skills having increasingly become a prerequisite for graduate designers and construction managers. During the recruitment process itself, the BIM Manager is often part of the interview panel posing questions to candidates in order to ascertain their proficiency in BIM and assess their qualifications. Inviting only candidates with a

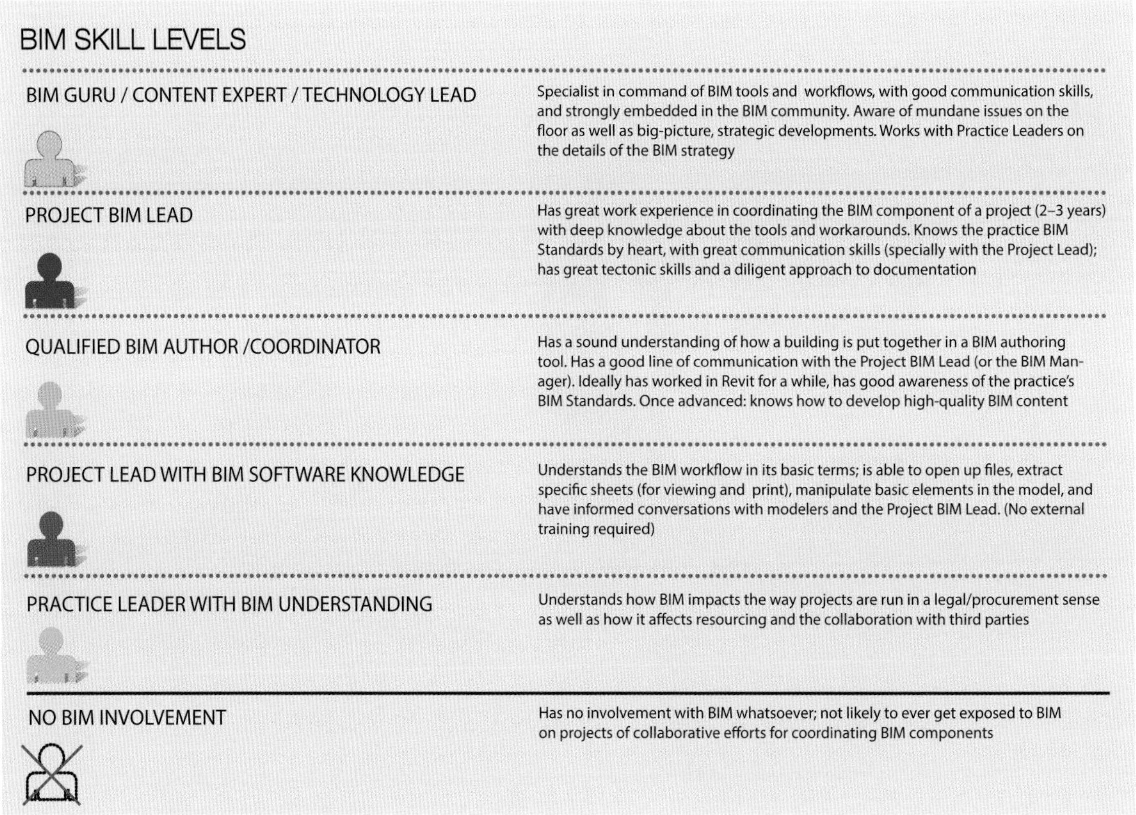

BIM SKILL LEVELS

BIM GURU / CONTENT EXPERT / TECHNOLOGY LEAD
Specialist in command of BIM tools and workflows, with good communication skills, and strongly embedded in the BIM community. Aware of mundane issues on the floor as well as big-picture, strategic developments. Works with Practice Leaders on the details of the BIM strategy

PROJECT BIM LEAD
Has great work experience in coordinating the BIM component of a project (2–3 years) with deep knowledge about the tools and workarounds. Knows the practice BIM Standards by heart, with great communication skills (specially with the Project Lead); has great tectonic skills and a diligent approach to documentation

QUALIFIED BIM AUTHOR /COORDINATOR
Has a sound understanding of how a building is put together in a BIM authoring tool. Has a good line of communication with the Project BIM Lead (or the BIM Manager). Ideally has worked in Revit for a while, has good awareness of the practice's BIM Standards. Once advanced: knows how to develop high-quality BIM content

PROJECT LEAD WITH BIM SOFTWARE KNOWLEDGE
Understands the BIM workflow in its basic terms; is able to open up files, extract specific sheets (for viewing and print), manipulate basic elements in the model, and have informed conversations with modelers and the Project BIM Lead. (No external training required)

PRACTICE LEADER WITH BIM UNDERSTANDING
Understands how BIM impacts the way projects are run in a legal/procurement sense as well as how it affects resourcing and the collaboration with third parties

NO BIM INVOLVEMENT
Has no involvement with BIM whatsoever; not likely to ever get exposed to BIM on projects of collaborative efforts for coordinating BIM components

Figure 5–6 BIM role description breakdown.
Copyright © Dominik Holzer/AEC Connect

those who model to interrupt their work at the point of handing it over for coordination, and only accommodate changes once the coordination meeting has been conducted. Otherwise stakeholders run the risk of coordinating model iterations that are out of sync. Equally, it is important that the BIM Coordinator detects any outdated model updates (or updates that don't pick up on all previously suggested changes by any contributing party). In those cases where individual contributions to the federated model appear to be outdated, the BIM Coordinator needs to flag this issue with the original author in an attempt to search for a solution. With progressing uptake of BIM collaboration via the Cloud and the resulting increase in connectivity among various stakeholders, there is a chance that in the near future coordination may occur in ever-shorter intervals. These intervals may get reduced to a point where federated models are accessed "live" and conflicts get resolved close to "real time."

Naturally, parties contributing to the combined modeling effort can also interface their efforts outside the prescribed coordination rhythm. For example, a mechanical contractor may wish to coordinate his or her design with the hydraulics contractor on a biweekly basis for detailed coordination. There is no issue with such individual approaches as long as they don't stand in conflict with the team's overarching coordination framework. Efforts in detecting clashes most likely reveal a great number of inconsistencies at the outset of coordination. The number of such clashes can go toward the thousands. Not all "clashes" detected by the software count as such. With the guidance of a BIM Coordinator, any multidisciplinary project team can soon identify and distinguish key points of conflict that need resolving. In most cases this will result in one party moving elements within their respective models in order to avoid spatial overlap with any other items. BIM coordination software such as Solibri™, Autodesk Navisworks™, and Tekla's BIMsight™ have built-in reporting functions that allow teams to identify those issues quickly with mark-up functions that juxtapose text-based entry (change suggestions) with visual feedback about problem areas.

With the coordination work progressing in the lead-up to construction, model inconsistencies within the federated model get gradually eliminated by the team. Due to the fast-tracked nature of many construction projects, there is a chance that the multidisciplinary team cannot afford to resolve all clashes prior to the commencement of construction. Such an approach—even if desirable—is not necessarily required. Instead it is common on projects that BIM Coordinators ensure that those areas are clash-free in the virtual (federated) models that are getting built first. In that sense, clash detection can focus on certain floors/zones, as long as any relocation of systems/elements does not fundamentally affect the layout in those other zones and floors.

Being a well-established process for illuminating inconsistencies in construction documentation and shop drawings BIM Coordination is becoming the standard method for design and construction coordination. Errors (e.g., clashes in the federated models) are easily detectable, and the BIM Coordination process is likely resulting in a high-quality "product" where problems get resolved on screen before they even occur on site.

Beyond the Model

BIM Managers and Coordinators working with construction companies have the opportunity to take the federated model further, interrogate quantities and cost, as well as link model content to construction timelines.

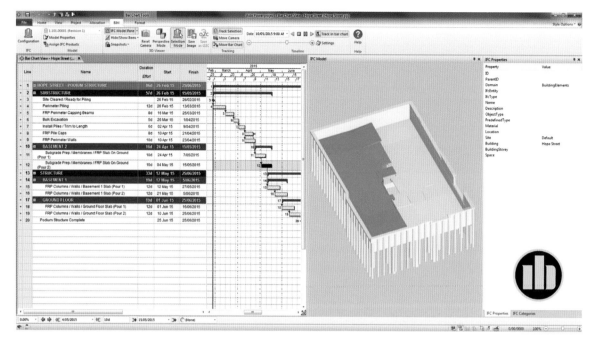

Figure 5–15 5D BIM cost and 4D BIM timeline.
Copyright © Produced by Mitchell Brandtman

A number of software options allow Project Managers and BIM Coordinators to link modeled content and Gantt charts for model-based construction programming. Flow-line scheduling based on BIM is becoming common-place on many jobsites and it can be associated with safety-on-site simulations. Construction Superintendents and Foremen then use this information to manage trades, resolve scheduling conflicts, record material movement, and to cross-check construction progress overall. 4D scheduling in BIM can even be linked to crane movement as well as other logistic activities on site.

Model data stemming from BIM can fulfill several more purposes on site. Next to increasing the builders' under-standing of spatial configurations of the systems/construction elements to be installed, BIM Models also serve the purpose of high-precision component setout on site. Formwork, walls, hangers, plinths, and others can all be positioned precisely on site via laser projection of 3D points taken from the BIM construction model. The Field BIM equipment used for this purpose (such as Trimble's Total Station) can also assist in recording the con-struction and installation progress, and feeding this information back to the model server via a Wifi connection. There it can interface with tools such as Vico's Office Production Controller™ that interface data stemming from the field with the overall construction schedule. Project Managers can use that information to confirm the exact status of construction progress, explore alternative scenarios in case of delays, and offer forecasting functions for future production.

It is a prerequisite for the application of BIM-enabled setout on site that the team sets up a Common Control Network—a collection of reference points (usually placed on primary and highly visible structural elements)—that guarantee accurate positioning and cross-referencing between BIM data and exact locations on site.

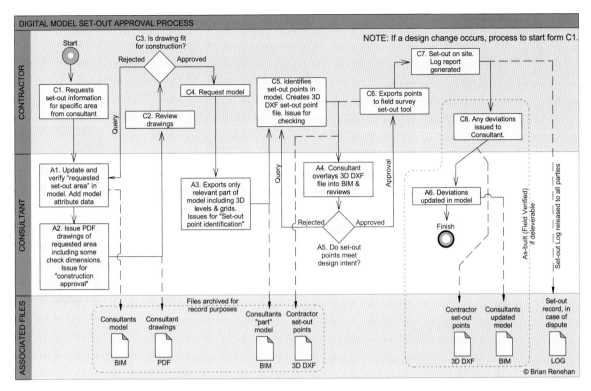

Figure 5–16 Digital setout flowchart proposal.

Copyright © Brian Renehan—BIMfix.Blogspot.com

BIMfix's Brian Renehan describes potential challenges in the handover of design data for component setout based on his own experience.[1] Renehan explains the digital setout process while also commenting on the risks of model handover from consultants to contractors. In the end, the benefits outweigh the risks if clear protocols are followed that ensure data integrity and checks by the relevant parties within their area of responsibility. Acknowledging that the main benefits are on the contractor side, there are also a number of advantages for designers who pass on their model data. In particular three aspects stand out from Renehan's summary:

- A reduction in last-minute "request for information" (RFI) queries,

- Complex designs are constructible, and the

- Method promotes a reduction of cost over-runs; thus the client perceives a greater service provided by the design team.

BIM Ping Pong with the Client

Clients are often not yet aware of the underlying dynamics inherent to BIM, easily misjudging what value they could ultimately get out of the BIM process facilitated by their project team. This kind of misinterpretation is more likely to occur if they have had little exposure to BIM projects and they apply traditional contracts to procure projects, set up teams, and determine the associated fee structure. The UK model of clients delivering

Figure 5–17 Data extraction schemer for generating FM information from BIM.

Copyright © Dominik Holzer/AEC Connect

TYPE PROPERTIES	MECHANICAL						PLUMBING					ELECTRICAL					
	Chillers	Boilers	Pumps	Air Handling Units	Fan Coil Units	VAV Handlers	Fittings & Fixtures	Pipework	Stormwater Drainage	Pumps	Tanks	MSB's	DSB's	Power Meters	Light Fittings	Motion Detectors	Dimmer Racks
#XYZ_Name	1		1	1	1	1	1	1	1	1	1	1	1	1	1	1	1
#XYZ_Category								1		1	1						
#XYZ_Description								1		1	1						
#XYZ_AssetType	1		1	1	1	1		1	1	1	1	1	1	1	1	1	1
#XYZ_ManufacturerName	1		1	1	1	1		1	1	1	1	1	1	1	1	1	1
#XYZ_ModelNumber	1		1	1	1	1		1	1	1	1	1	1	1	1	1	1
#XYZ_WtyGuarantorParts								1	1		1	1					
#XYZ_WtyDurationParts								1	1		1	1					
#XYZ_WtyGuarantorLabor	1		1	1	1	1		1	1	1	1	1	1	1	1	1	1
#XYZ_WtyDurationLabor	1		1	1	1	1		1	1	1	1	1	1	1	1	1	1
#XYZ_WtyDurationUnit	1		1	1	1	1		1	1	1	1	1	1	1	1	1	1
#XYZ_ReplacementCost								1	1	1	1	1					
#XYZ_ExpectedLife						1		1	1	1	1	1					
#XYZ_WtyDescription						1		1	1	1	1	1					
#XYZ_ModelReference						1		1	1	1	1	1					
#XYZ_ElecConsumption								1	1	1	1						
#XYZ_WaterConsumption				1				1									
#XYZ_GasConsumption			1	1				1									
#XYZ_NominalLength			1	1				1	1	1							
#XYZ_NominalWidth								1	1	1	1						
#XYZ_NominalHeight								1	1	1	1						
#XYZ_LoadLimit			1	1		1		1		1	1	1			1		1
#XYZ_Size																	
#XYZ_Capacity	1		1	1		1		1	1	1	1	1	1	1	1	1	1
#XYZ_Color																	
#XYZ_Finish																	
#XYZ_Material																	
#XYZ_Location			1	1	1	1		1	1		1	1	1	1	1	1	1
#XYZ_SAPAssetCode			1	1	1	1		1	1		1						

their Employers Information Requirements (EIR) to the project team is a promising approach to minimize the schism between client expectations and the services offered by the team. The problem remains that clients are still often not familiar with expressing their requirements in BIM terms. With a lack of knowledge what BIM can offer them, they need support in defining their deliverables. This is where the BIM Managers come in to highlight the opportunities.

One way of doing so is to start a dialogue as part of the return-brief where there is a give-and-take between clients and the project team: On one side clients and their Facility Management representatives explain their in-house processes; on the other side the BIM Consultants highlight potential benefits of BIM for the client. BIM Managers can expect to be asked how each of the additional services provided are estimated. Based on this feedback clients conduct a cost-benefit analysis to determine the level of BIM that's right on the project. There are likely a number of interactive sessions required to fine-tune what is possible with what is commercially desired (Ping Pong). It also makes sense for BIM Managers to schedule a number of client-focused workshops to advance the BIM deliverables together with the client's Facility Management team. In those sessions BIM Managers then determine the preferred output format for any data stemming from BIM to be handed over to the client. With COBie (explained in Chapter 3, "Focus on Technology") being one option, there are other formats and approaches that may be considered. Key to the success to such workshops and the resulting BIM-FM handover is to increase the understanding what the client plans to do with the information that can be extracted from BIM. In this way the kind of information to be handed over and the exact attributes that clients want associated to geometric models can be determined. The workshops should also determine the timing of BIM-FM data-drops so BIM Managers can weave this information into the BIM Execution Plan.

BIM Managers need to be careful to avoid reflecting merely on those deliverables in BIM that get produced by them directly; instead they need to emphasize with the core activities of the client to extract key points of benefit for their business. For example, the benefit of a well-coordinated design that avoids issues on the construction side is only a minor selling point for the client. They expect such service as a matter of course and are likely not going to see it as a value-add. Simultaneously highlighting the benefits for construction coordination via virtual interfaces has little extra value for clients who are more interested in the O&M side of things. A short animation that highlights the construction process and timeline may be beneficial though as clients can use it for their marketing and publicity.

Fire-Fighting and Lending a Helping Hand

Whereas the above sections did highlight what a BIM Manager can do in order to approach project support in a structured way, there often remains an ad-hoc component to project-based BIM Management that seems unavoidable. There exists a paradox situation where the provision of immediate support on projects is a highly rewarding experience for a BIM Manager, whereas at the same time it is potentially the least productive out of all his or her tasks.

With all the opportunities for BIM Managers to set up a support infrastructure described in this publication, there will always remain a component where BIM Managers will be asked to provide immediate on-project support. In some cases this will be in the form of walk-ups from colleagues who simply have a software-related question; in other cases BIM Managers may be asked to compensate for a shortage of manpower on a project. Other cases again will see BIM Managers searching for workarounds and quick fixes that allow the project team to continue their work undisputedly.

Such fire-fighting exercises occur frequently for a broad range of BIM- and IT-related issues. One should not confuse common walk-ups of staff who have a specific "how-to" question with "issues that cannot be anticipated." BIM Managers at times see pride in being able to provide ad-hoc support, thereby reacting to an immediate need of a colleague. Such an effort though should come with a warning: As much as this form of support is sometimes unavoidable, it is highly counterproductive: The person receiving support only seldom benefits from it beyond a quick fix. He or she is unlikely to learn how to deal with a similar issue the next time. An overly responsive BIM Manager risks being instrumentalized and taken for granted as the go-to person whenever an issue arises. Probably even more problematically: Being caught up in endless on-project mentoring can take over the majority of a BIM Manager's time, thereby detracting him or her from his or her core duty—Managing!

As discussed earlier in this chapter, one way to avoid this marginalization is to clarify the proportion of project- and non-project-related work in a well-written business plan. This distinction is relevant for a number of reasons, one of which relates directly to assigning BIM to the project fee, or as a general cost that cannot be

Figure 5–18 BIM Manager assisting delivery on the floor.
Copyright © Point Advisory

with others to advance what they are doing and in some cases assist in expanding where BIM is headed locally and globally? The following responses by four global BIM leaders reflect on the key sources of information BIM Managers rely on to advance their skills outside of their standard training. The nature of engagement presented here highlights the relevance of proper planning, alignment of BIM goals with the overarching goals of an organization, and social engagement with others and more as the essential ingredients to achieving BIM excellence.

OUR GLOBAL BIM LEADERS' TAKE ON WHAT MAKES A BIM MANAGER EXCEL

Ronan Collins: Excellent communication skills—the ability to listen, discuss, and explain ideas—are crucial to being a BIM Manager. Whether you are operating within one organization or acting as the BIM Manager for a project, you need to enroll everyone, getting them involved in the BIM process. This takes a concerted effort, a lot of dialogue, and an ability to be flexible in considering other people's opinions; balanced with an assertive approach when needed.

Rob Jackson: The single most important thing that a BIM Manager must understand is the process for the particular business he or she supports. Without understanding this, it is difficult to introduce change as users will simply fight back with "you don't understand the process." Most BIM Managers I have seen are organized, thorough in their approaches, have some tendencies toward OCD, determined and have ability to think outside the box. The role is one of both working in isolation but also being part of a wider team. There are periods where development needs to take place and you simply need to get your head down, but at the same time talking to users about emerging ideas and getting buy-in is also critical for the point when changes are made for real.

A good BIM Manager stands by their proposals but also takes on board feedback to improve the process.

Paul Nunn: From my point of view, it is experience in construction as an architect, engineer, or design manager that makes a BIM Manager excel. Apart from working as a Contractor's BIM Manager, I also audit a lot of BIM projects for clients and contractors and continuously come across graduate architects or engineers or CAD managers who have never been involved on site or with a contractor and quite simply don't understand the construction delivery process.

Becoming a BIM Expert

"Personally I believe that there are no 'BIM experts.' Can one person really understand every requirement of the workflow of design, construction, and operation? The reality is the subject matter is too great for any individual to truly grasp."

Rob Jackson, Associate Director at Bond Bryan Architects

Can one speak of BIM experts? If so, how does one become one and how does one remain on top of the game in a context that is constantly expanding and changing?

BIM expert status is not gained overnight. It is usually derived out of a combination of getting one's hands dirty in daily delivery, and by following-up on the latest developments outside of the project realm. When immersed

Figure 6–3 Cathay Pacific cargo terminal.
Copyright © Intelibuild

in projects, it takes a good number of years to realize any real level expertise in the field. In doing so, BIM Managers will likely follow a particular path related to their specific domain, whether it is architecture, engineering, or construction. Fundamental to getting peer and industry recognition for a particular area of knowledge is an understanding of its limits in terms of application, as well as the level of impact that expertise has on advancing a project overall.

Next to experience gathered on projects, BIM Managers advance their skills by attending specialized training courses and workshops, engaging with BIM blogs, mingling with fellow experts in local user-group meetings (traditionally based on individual software use but more frequently cutting across a range of BIM applications), attending and presenting at BIM conferences, active participation in discussions on specialist groups within professional networks such as LinkedIn, and so forth. Typical additions to these points are "out of work-hours" sessions where BIM Managers catch up on the latest software developments, useful workarounds, and additional self-training. In other words, BIM Managers constantly have to stay up to date with the latest developments. Technology is one factor that is changing rapidly, but BIM Managers also need to remain alert to any news related to process and policy changes surrounding BIM and Design Technology.

Robert Yori: Expert status can best be described as how well one can apply technology to practice. I'm fond of making comparisons to language—while one may become an academic expert in linguistics and various languages, that knowledge is most effective when used to effectively communicate one's ideas. The same could be said for BIM. Procedural and software knowledge must be understood at visionary and strategic levels, and then executed in tactical ways that lead to project successes.

Understanding the AEC culture and localized environments are also essential. BIM can promise some pretty drastic changes, and in order to be effective at any type of change management one must make the effort to understand the status quo, and when possible, why it is so. Change and compromise begins with dialogue, and dialogue begins with understanding and respect.

Rob Jackson: Expert status is more likely to be gained in a specific area or a few areas of BIM, which a BIM Manager has some level of expertise in. In many ways an expert is one who knows when and who to ask. The expert status comes from building up a network of other individuals who can assist a BIM Manager to find the right answers.

Paul Nunn: It is hard to be an expert. I think it's more about having a strong general understanding of the design and construction process, sufficient so you know what questions to ask, who to pose the questions to, and how to frame those questions.

Ronan Collins: While I agree with the sentiments of the others, I believe individuals can become experts within their own domain. For example, a BIM Manager working within an organization, say an architectural practice, could become an expert BIM Manager for the design team. He or she should have a degree, have worked for at least five to eight years in the industry, and be experienced in producing the models, drawings, and schedules in collaboration with other design consultants. He or she also needs to be aware of the needs of the contractor so that they set up projects appropriately. Expert BIM Managers for contractors need to understand project scheduling, how to manage deliverables by trade contractors, and how to use BIM for cost control. They also need to understand how the designers have approached the design phase and when possible, give direction to the designers to make the design models more effective.

Innovating with BIM and Educating Others

"The challenge is not in finding shiny new toys but in understanding how they apply to an organization or project, if at all."

Ronan Collins, Managing Director at InteliBuild

As a BIM Manager, it is important to be able to translate one's knowledge about proven best practice into a set of instructions that outline typical workflows and BIM-related processes to others. Such knowledge is not merely gained via project experience, but it often also draws on the experience of others who are able to distill how they mastered a potential issue and who consequently provide concise descriptions to a wider audience

Figure 6–4 BIM review.
Copyright © Intelibuild

on how that was accomplished. Over the past few years, social media has played an important role in supporting BIM innovation among a global group of users. BIM Managers need to put time aside to remain on top of relevant developments in their field. Among the various types of media used, BIM blogs are one crucial way of sharing knowledge with others.

The BIM Blog

"Social media has been invaluable in terms of keeping up to date with current articles, documents, and standards. Social media has also introduced me to other users and 'experts'."

Rob Jackson, Associate Director at Bond Bryan Architects

BIM blogs have become a highly regarded source of information for BIM Managers who skim through their content on a regular basis to learn about the latest updates to software development and other in-depth insights from trusted industry experts. There are principally two types of blogs. One type is predominantly written by a single person who posts more or less regular updates that range from weekly to monthly or even longer intervals.

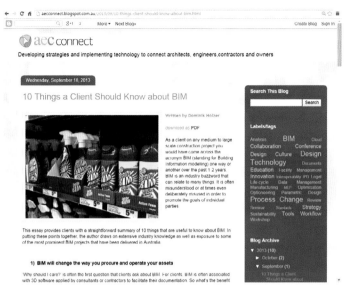

Figure 6–5 BIM/Design Technology blog example.
Copyright © Dominik Holzer/AEC Connect

At times these experts invite guest bloggers to provide their insights. The second type is a blog with a larger number of contributors. In that case it is common for the blog to be updated at least weekly, if not daily. Many of these blogs have become highly popular among BIM experts. They are covering a broad range of BIM-related aspects from strategic to practical, from software-focused to process-centric. Notable early examples (some of which are still updated regularly) include: The Laiserin Letter,[1] Lachmi Khemlani's AEC Bytes,[2] Paul Wilkinson's Extranet Evolution,[3] James Van's All Things BIM,[4] and many others. More recent examples of BIM blogs include Laura Handler (bim)x—now under her own name,[5] Spacework's bimcrunch[6] (with daily BIM news), Antony McPhee's Practical BIM,[7] James L. Salomon's Collaborative Construction Blog,[8] Randy Deutsch's Data Driven Design,[9] Rob Jackson's BondBryanBIM,[10] or Brian Renehan's BIMfix.[11] There also exist a number of software-specific blogs, most notably from the Autodesk User Group International (AUGI)[12] or blogs such as Revitall,[13] What Revit Wants, the BE Usergroup blog, the Archicad blog,[14] ArchiCAD monkey,[15] Nathan Miller's The Proving Ground,[16] and many more. National and International BIM proponents such as the buildingSmart group also have strong presence online with blogs and support websites such as http://www.buildingsmart.org/ and http://open.bimreal.com/bim/.

RUNNING A BIM BLOG

1. Do your research on what is already out there. What do you have to say that could add to this?

2. Be aware that a blog should be interactive, write in a way that invites others to respond.

3. Apply a writing style that is concise and to the point.

4. Include images and/or videos to make your material more engaging and explicatory.

5. Publicize your blog entries in professional online fora and networks such as LinkedIn.

6. Be quick to respond to any answers to a blog post; start to curate a conversation and allow for different points of views to be expressed. Fostering a lively discussion about a particular BIM-related topic will ensure the relevance of your contribution.

A recently added online resource that focuses predominantly on BIM videos is the B1M,[17] set up by Fred Mills and Tom Payne. Here members can gain access to expert advice in a video format, often produced by peers who are willing to share their insights.

Subscription and membership to these blogs is usually free and those who subscribe can be sure to receive notifications as soon as a new post gets published. Some bloggers have chosen to make their archive material available at a cost, thereby supporting their ongoing hosting costs. For many, blogging is not only a way to share their knowledge with others, but also to raise awareness within the industry of the in-depth knowledge of those who post—in particular if their views are shared with a larger audience (nowadays blog posts can easily go viral when tweeted by renowned opinion leaders). Creating a worthwhile and original BIM blog takes time and dedication. In terms of credible profile building BIM blogs clearly underline an individual's expert status and they become like an online portfolio for any BIM Manager or technology expert who runs them. Any BIM Manager who decides to start a blog should be aware what is out there already and how he or she can build form and add to the existing body of knowledge. Looking for distinction is key to success for an outstanding BIM blog, either via providing rich and novel insights that get shared, or simply via rigorous testing that can benefit a greater community. Once a blogger has established a strong credible foundation, he or she is likely to be able to build up a great community of subscribers and re-posters. Outstanding BIM blog posts are often linked to online professional networks such as Linkedin™. A lot of in-depth content on blogs is also constantly being referred to via other social media BIM protagonists via Twitter™.

Changing People's Minds about BIM

Responding to issues such as innovation and overcoming change resistance, the four industry experts who were invited to comment on this section provide the following responses:

THE EXPERTS' VIEW ON IMPLEMENTING INNOVATION AND OVERCOMING RESISTANCE TO CHANGE

Rob Jackson: BIM Managers need to be given time to research and test new ideas. The key to overcoming change resistance is demonstrating workflows that are simple to explain and straightforward for a user to implement. If you ask 10 users to implement a new workflow probably two or three will get it straightaway and they will go off and implement it immediately. This process then gives a BIM Manager further evidence that the workflow can be implemented on live projects. It provides evidence internally to staff, but it can also be used for the creation of marketing materials. The implementation process is then used to support prequalification scenarios as it's better to show real examples than simply theoretical workflows.

Of course the early adopters of new workflows will also provide further feedback and in return allow workflows to be finessed further. The second wave will adopt the tweaked approach having been convinced by their peers. Before long the percentage of users adopting the workflow will reach a critical mass and it then won't take long for the remaining users to change their approach.

I would also say that the most successful changes are the ones staff can't avoid or provide real benefit to them. For example, we integrated some of the required COBie fields onto our drawing sheets. These fields also go a long way to structuring our IFC models for other use. Users still care and focus on drawing output but by adding the fields to the standard drawing sheet there was no way to avoid filling out this data. Some didn't realize this was for COBie. This is the ideal scenario where small changes can create other benefits.

Paul Nunn: There needs to be sufficient time to do the research, attend the conferences, and participate in the relevant webinars. This partly comes down to getting fees good enough to enable additional staff or time off to do this research. Change management in our own organization isn't a problem. However, change management is still probably our biggest problem with most contractors we work with.

Ronan Collins: From experience, the current generation of BIM Managers is very aware of lots of different innovations in technology both for BIM and the broader industry, including cloud computing, big data, laser scanning, 3D printing, RFID and site setting out tools. As a BIM Manager, you have to evaluate an existing practice, determine if an innovative idea can improve your team's productivity, or provide value to your customer. Only then can you work out how to implement the changes needed and engage users in changing their ways.

Robert Yori: The best way to effect change is to demonstrate something that's compelling enough for people to want to emulate or investigate it. In the context of BIM, it might be a project that was able to run with fewer staff, or one that produced a minimal number of Requests For Information (RFI)s or, a narrow range of bids. It might have made the relationship with project consultants easier or more fulfilling. When firm leaders and employees can begin to understand the benefits BIM can bring to a project in terms that directly affect their involvement, BIM resonates.

BIM Research

"Build on the successes and lessons of others, and share yours so that others may build on them."

Robert Yori, Senior Digital Design Manager at SOM

One of the most striking aspects of the way BIM advances globally is the effort put into its ongoing research and development in practice. One would typically associate research activities related to technology and tool development predominantly to an academic, or a software-vendor-specific setting. As much as this is the case to a degree in the context of BIM, the proportion of BIM research occurring in practice is particularly high. BIM Managers are a driving force behind the advancement of BIM, not only in terms of technology, but also in terms

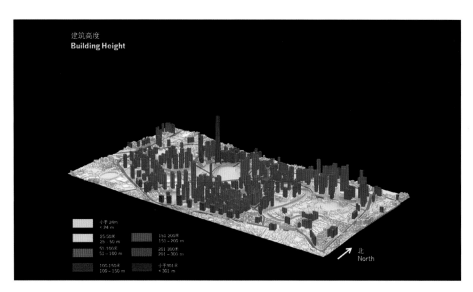

Figure 6–6 City information modeling.
Copyright © Skidmore, Owings & Merrill, LLP

of process and collaboration. Whereas the user-group meetings, conferences, and blogs discussed previously are a pivotal part to advance the discourse about BIM, focused research as it occurs in practice is taking it to the next level. For some it is only a small step from showing engagement with BIM by commenting publicly to actually expanding its scope via dedicated Research and Development (R&D). In those cases where research efforts related to BIM are of a mainly technical nature BIM Managers and/or their IT colleagues search to expand on the "off the shelf" capability of their tool by writing add-in scripts and functions that benefit their particular workflow. BIM R&D may relate to developing simple add-ons all the way to facilitating major additions that rely on access to a BIM authoring tool's Application Programming Interface (API). In many instances, well-conceived plugins offer modelers additional options to fulfill their tasks more quickly and better on projects in-house. In other cases, tool development can deliberately result in the generation of applications that provide potential value-add to a broader community. In those cases it is not uncommon for startups to develop their business by targeting specific BIM goals/tasks that off-the-shelf software does not provide. Some may continuously look for BIM-related functions desired by users but not offered as standard by existing software solutions. Others may set up their business in the hope to get "acquired" by a larger software seller who subsequently incorporates their solution to expand the standard functions of their tool.

BIM Managers who push for the commercialization of their research output should better think twice. Not only do they run the risk of straying away from supporting their organization's core business (which is unlikely going to be software development), they also need to consider the added effort of generating a product or service stemming from their BIM research. The key difference is the need to ensure robustness of the product which is usually facilitated via time-consuming de-bugging and testing. Further, they need to consider providing support to those who purchase their tool and update it on a regular basis.

The more common form of value-add of BIM research typically stems from focusing on value-add to the BIM Manager's firm or its clients. For any research to be undertaken, BIM Managers are therefore advised to consider key selling points of what they develop beyond simple practicalities of facilitating the office-internal workflow. Where there is pressure for the research to pay for itself, the output will have to result in something that can be marketed as a "value-add" to the client. In many cases this may relate to a particular way they can engage with their projects virtually—either via visual interrogation of the design outcome, or via data sources that interact with their business planning and validation.

Larger BIM software developers often depend on a group of core users to help them advance their products. Such advancement can occur in different ways. BIM Managers who can prove their organization's track record in testing and advancing BIM software may push to be included as a Beta tester. Once accepted within this exclusive club, they will be given access to new versions of software before their official release. By feeding back their experience on testing the tool they actively help advance it, adding new features, and de-bugging any remaining inconsistencies in the software. Beta testers usually exchange updates in dedicated groups such as the Autodesk Developer's Network (AND) or Bentley's "Betas" committee. Access to such groups usually requires members to go through a strict application procedure and to sign a confidentiality/nondisclosure agreement. In some cases, membership may come at a cost.

HOW DO YOU GO ABOUT BOOSTING YOUR FIRM'S BIM CAPABILITIES?

Rob Jackson: In the United Kingdom the most important aspects to understand are the standards and protocols that have been developed as well as the technology we use. However, I have also learned to challenge everything I read and hear and work out what is relevant to us. This includes the information surrounding the government mandates as well as general BIM-related information that I come across in writing and when attending presentations.

Paul Nunn: Keeping up to date on the various approaches to BIM around the world is key to a firm's BIM capability. There are now a lot of countries and states mandating BIM, as well as a lot of different organizations and much of their BIM approach is publicly available on the web. Each of these may have one significant new idea that can be adapted to your in-house approach.

Ronan Collins: This topic should be considered in light one of the fundamental weaknesses of our collective design-build-operate industry. Construction-related firms do not invest time, energy, or money in research and more importantly development of ideas or of staff. The low cost, highly competitive tendering process we all participate in, leaves no surplus for research or training. As a BIM Manager you are expected to have the answers, to have processes that are tried and tested, and to implement tools that work "out of the box." This is probably the most daunting part of a BIM Manager's role. To overcome this, you can join local associations, attend seminars, and participate in online forums. The best research tool available is to ask your peers for their advice directly and be willing to share your own solutions with others.

Robert Yori: Always be on the lookout for tips and techniques. Make sure to be plugged in to the BIM and tech communities, local and beyond. Keep an eye on what's happening in academia, and what capabilities new graduates are bringing to your firm. And if you think there's a better way to do something, try it.

Reaching Out and Getting Noticed!

"Meeting individuals either in passing at events or at more formal meetings has allowed me to explain our approach. This is as much about showing them what we do as getting their feedback and suggestions for improvement. This type of feedback also comes from undertaking presentations."

Rob Jackson, Associate Director at Bond Bryan Architects

Earlier in this chapter the relevance of BIM blogs and social media, which connects BIM experts worldwide with their peers, was discussed. This section looks specifically at how face-to-face meetings and attendance at events might complement a BIM Manager's online presence, whether it is through local user groups or BIM Technology/Strategy/Management conferences. Networking in BIM circles in person provides a double opportunity: to meet others and learn from them, while also showing others what one has done in order to add to the BIM discourse.

Meeting Others Locally—BIM User Groups and More

As much as online presence is a welcome source of information for BIM Managers to advance their skills, meeting others in person remains an essential component of advancing one's skills, sharing views with others and expanding one's network. Methods for encounter range from small user-group meetings organized by a team of local BIM aficionados to larger public BIM fora.

Whereas traditionally CAD user groups would predominantly occur in isolation within singular professions, BIM user groups open up new channels of communication to colleagues from other disciplines. It is common to find a great mix of stakeholders at BIM-related events. One notable inclusion is the attendance of upper management on BIM-related sessions, as dedicated BIM sessions on Construction Law, Project Management, and Facilities Management become more frequent. Ideally, these events trigger response from a broad stakeholder group who get together to discuss challenges and pain points that affect progress of BIM across the industry. The UK government's BIM Task Group has undertaken great efforts in recent years to connect industry

Figure 6–7 Presenter at MelBIM, Australia.

stakeholders; they work collaboratively with other professional and industry organizations, such as the Building Research Establishment (BRE), Royal Institute of British Architects (RIBA), Royal Institute of Chartered Surveyors (RICS), and the National Building Specification (NBS), on joint BIM events. In the United States, local user groups, such as the New York City Revit User Group (NYC-RUG),[18] attract professionals from diverse backgrounds and debate the latest and greatest in BIM. In Australia BrisBIM,[19] serving the Brisbane region, kicked off a trend for regular trans-software, trans-discipline BIM gatherings in the state of Queensland by attracting visitors by the hundreds. This has since been replicated in Melbourne (MelBIM[20]), Sydney (SydBIM[21]), and across Western Australia (BIMWest[22]). These events clearly build a community of like-minded people who gather in a relaxed social environment to discuss BIM outside the pressure of everyday project deliverables. BIM fora not only attract local BIM Managers; they become a platform for encountering those who simply want to learn more about how BIM affects their daily work both in terms of technical aspects and in terms of strategic planning, policy, and business drivers associated to BIM. Often, local BIM events are run by a group of enthusiasts who always search for captivating proposals from potential speakers. Any BIM Manager who feels they have a valuable contribution to make should approach the speaker managers of their local event with a proposition for a presentation. At times this will result in presenting lessons learned from a specific project they have worked on (in this case, one has to ensure permission by the client and other collaborators to share project-specific information). Alternatively, presentations to the local BIM group can revolve around a more overarching issue. Either way, it is relevant for any presenter to introduce themselves and their organization appropriately at the outset of any talk, but without too explicitly promoting their own organization. The best publicity stems from the quality of the material presented and the way one masters certain challenges (often in collaboration with others).

BIM Conferences

One step up in scale and relevance from regular local BIM user-group events are BIM conferences. Ever since the increase of BIM use in the mid-2000s a number of such (often annual) events have emerged. In some cases they are organized by software developers as a means to promote the capabilities of their software (not just BIM) within a broader context of BIM and Design Technology adoption. Examples of such events include Bentley's "BE Together" (now part of Bentley Learning[23]), Autodesk University (AU[24]), Tekla Conference,[25] just to name a few. These events usually attract an audience in the hundreds, if not thousands, and the presentations usually include a mix between software updates offered by the vendors and associated tool-makers, as well as papers presented by BIM experts from around the world. These events also typically include BIM awards where cutting-edge project work gets assessed by a jury and prizes for the best work in various categories are handed out. An emerging trend among these conferences is to franchise them out into smaller local happenings where vendors run information sessions about new software releases that then get embedded into a broader discourse about BIM.

Next to these events which always have an element of a "tradeshow," BIM conferences also regularly get organized by industry bodies and groups. There the focus typically lies less on software itself, but on the impact of technology such as BIM on professional development and practice overall. A trend has started over the past few years where industry bodies and groups such as buildingSmart increasingly push for themes and presentations

Figure 6–8 BIM-MEP^Aus forum.
Copyright © Dominik Holzer/AEC Connect

that cut across disciplines instead of focusing merely on what BIM might mean for any single profession. As seen in the United Kingdom, some industry bodies join forces in offering BIM conferences that synergize different aspects of project delivery into a single event, thereby attracting a broad audience who are given solid insights from a lifecycle BIM perspective. One notable Australian example incudes the BIM-MEP^Aus Forum[26] by the Australian Mechanical Contractor Association (AMCA). This annual event has managed to attract a global audience with speakers from property, construction, engineering, and architecture with their focus on information handover between various stakeholders who use BIM.

Another type of gathering that has gained strong momentum in recent years is the BIM User Conference. In contrast to events organized directly by software developers, these conferences tend to be run by independent organizations with close ties to BIM user communities. Out of these, the Revit Technology Conference (RTC)[27] has seen the greatest success since its inauguration in 2005. Starting in Australasia, the RTC concept has since been expanded to local North American, Asian, and European events. What sets this type of event apart from others is the highly collegial nature of the peer-to-peer support provided during the lectures, classes, and hands-on lab sessions. RTC's Wesley Benn explains the reason behind the success of user conferences as follows:

"An unintended consequence of the increasing technical sophistication of our tools and our industries has been isolation, and the lack of personal connections that can be so critical to understanding how someone has reached a particular result. Conferences, where learning can be mixed with significant networking opportunities, provide the environment to offset this problem, to bring team members together, and to share knowledge and ideas across a much broader community than would otherwise be possible."

Wesley Benn, Chairman RTC Events Management

(a)

(b)

Figures 6–9a-b Impressions from a Revit Technology Conference (RTC).
Copyright © RTC Events Management Ptg Ltd

In the early days, user conferences tended to be highly software specific, but they are now becoming ever more inclusive to the point of being software agnostic. User conferences are less orchestrated than vendor-initiated events and presenters are nearly exclusively peer BIM Manager/experts or, more broadly speaking, Design Technologists. At RTC, speakers are selected based on the quality and relevance of abstracts they submit to present a particular topic worth debating publicly. BIM Managers who want to be recognized as an opinion leader in their field should therefore be alert about topics worth discussing and identify gaps in what has been presented previously. This approach usually works well if a speaker is selective about a specific area related to BIM on which to focus. With BIM expanding into ever more areas related to design, engineering, manufacturing, construction, and operation, conference organizers typically classify presentations into different fields. At RTC these currently relate to: Architecture, BIM, Business Strategy and Leadership, Civil & Infrastructure, Coding, Content and Customization, Construction and Fabrication, Estimation, General (multidisciplines), MEP, Operations and Maintenance, Simulation & Analysis, Structure, Visualization, and more. In addition to these overarching topics, user conferences typically also search for cutting-edge lab sessions where experts run hands-on tutorials with the latest software (ad-ins) and beyond. The benefit of a user conference lies not only in the event itself, but also in the fact that presentation material is usually made available online by the organizers, thereby offering a valuable resource for BIM Managers to refer to beyond the conference itself. If the material is structured well and easily navigable/searchable, it becomes an online "best-practice BIM" repository.

Some of the most successful presentations emerge out of collaborations between two participants from different professional backgrounds. Explaining how BIM can help to overcome professional boundaries or simply how it changes the mode of engagement offers the audience great insights about its potential.

Anyone presenting at these user conferences should be able to empathize with their audience. An informed audience of fellow users is typically less impressed by a sales pitch about all the great things one has achieved using BIM on a project. Instead, they rather long for an honest account about what worked and what didn't (and if so, why it didn't). Lessons learned that can be referenced into other contexts are a key component of great presentations.

PREPARING FOR A BIM PRESENTATION/CONFERENCE PAPER

1. Stick to a particular topic that you know much about.

2. Check what has already been written/presented in that area—ensure your contribution fills a knowledge gap.

3. Start by raising a question: Why is it that …?

4. Ensure there is a narrative in your presentation/paper: problem, solution, path of getting there.

5. Promote yourself/your firm through the quality of your contribution, not through repeated advertising.

6. Avoid merely presenting what projects you have been involved in; instead focus on particular issues and express "lessons learned" that others can benefit from.

7. At the end, explain what topics you want to cover next—burning BIM issues that still require resolution.

Distinguishing Your Service Offering via BIM

"A BIM Manager can play their part in developing a firm's reputation in the market place. They do it best by making the BIM aspects fade into the background and allowing the designers or construction professionals to shine."

Ronan Collins, Managing Director at InteliBuild

Figure 6–10 BIM coordination.
Copyright © Intelibuild

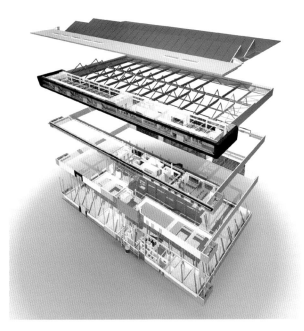

Figure 6–11 Nuclear Advanced Manufacturing Research Centre (NAMRC), Rotherham, UK.
Copyright © Bond Bryan Architects LTD

193   6: Excelling Your BIM Efforts

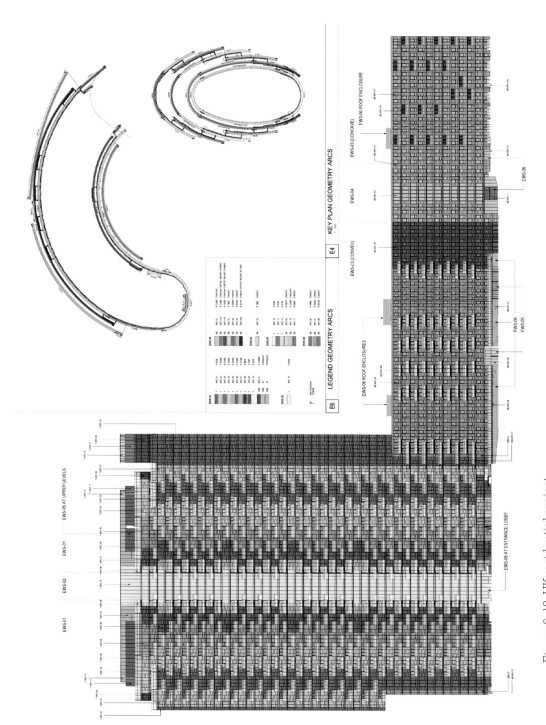

Figure 6–12 UK residential project.
Copyright © Skidmore, Owings & Merrill, LLP

One essential lesson to be learned by any BIM Manager who wants his or her firm to excel in BIM is to avoid trying to achieve too much. Distinction and excellence in procuring BIM can best be reached with a BIM strategy that focuses on supporting a confined set of activities. BIM Managers who claim to know everything about BIM and who attempt to market services related to an exhaustive range of BIM activities are likely to struggle enabling their organization to make an impact on the market via BIM. By tying specific BIM services to the core business within any organization and by offering the client value-add via associated information, BIM Managers can put their firm on the map and distinguish their offering from the competition.

Putting your firm and your own abilities on the map as a BIM Manager often relates to practical aspects of BIM work such as the delivery of high-quality documents and models that consider use by others downstream. Those BIM Managers who facilitate seamless coordination on projects enable the various contributors to work synergistically with well-formulated BIM Execution Plans and associated delivery processes. These qualities are highly regarded on the market both by the clients, as well as other stakeholders who form part of the collaborative process.

With off-the-shelf BIM becoming less of a distinguishing factor in the industry, BIM is still a pivotal approach to ensure client satisfaction and secure return business. Sometimes achieving such success comes from the facilitation of services that simply work on a practical level. Some of those may relate to opening up communication channels, the reduction of time-consuming and repetitive tasks, the link of useful information across stakeholders, and more.

Whatever the distinctive business a firm may be promoting via BIM, it is pivotal for them to state their value proposition and establish themselves as a capacity in their market locally and beyond.

HOW MIGHT A BIM MANAGER GO ABOUT PUTTING HIS OR HER FIRM ON THE MAP?

Robert Yori: It's not as easy as it once was. Years ago, all one had to do was to use BIM. Today, success lies in finding ways to resonate deeply within a firm's business goals. I am involved with the AIA Technology in Architectural Practice Group (TAP), and every year we give awards based on how well technology has been used on a project. This is exactly what we're discussing at the moment—now that everyone's "doing BIM," what are our criteria for the awards? It becomes less about making "the perfect model" and more about creating something that's solving problems encountered in practice. How are they helping to achieve what was previously considered impossible?

Rob Jackson: For me it's about picking a specific area of the process and developing a real expertise. We have developed our approach around open data-rich workflows but I have seen other architects focus on visualization techniques, environmental integration, use of mobile technology, and BIM to FM workflows. There is no right way to go, but the approach will be determined by a business's needs.

Paul Nunn: Our clear benefit is that our staff all come from a strong site-based contractor background and understand what a contractor wants. Another contributing factor is our attitude of being totally open to changing and adapting our processes to what our clients ask for. Again, during the BIM audits I do, I find a lot of BIM Managers have a standard Building Execution Plan (BEP) and just roll it out again and again.

Our in-house software development team also helps because we are very much focused on integrating models with existing 2D software such as Dassault's Enova™, Atlassian's Jira™, and various FM software.

Ronan Collins: BIM is a powerful enabler for better designs and more efficient building practices. The results of a well-planned and fully supported implementation BIM plan are what will make firms stand out from the crowd!

Embracing Lifecycle BIM

"Our approach has always been our focus on whole of project BIM."

Paul Nunn, General Manager BIM at PDC

With much focus required by individual businesses to carve out their particular expertise and position their firm on the market, BIM also requires a great understanding of processes for information handover. By now it is accepted that BIM can assist not only during design, engineering, and construction, but in managing the entire lifecycle of assets. The way in which information gets formatted and passed on from stakeholder to stakeholder requires the attention of the BIM Manager who often helps to drive project coordination. In the past, one detectable weakness within this process has been the gaps between Design and Construction BIM. Without this issue being properly resolved, much attention is currently given to the link between BIM and FM to satisfy the owner/operators' needs.

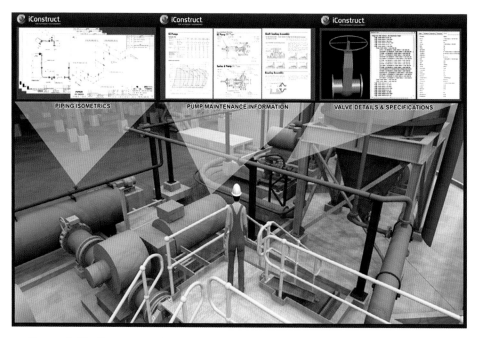

Figure 6–13 iConstruct information interfaces.
Copyright © PDC Group

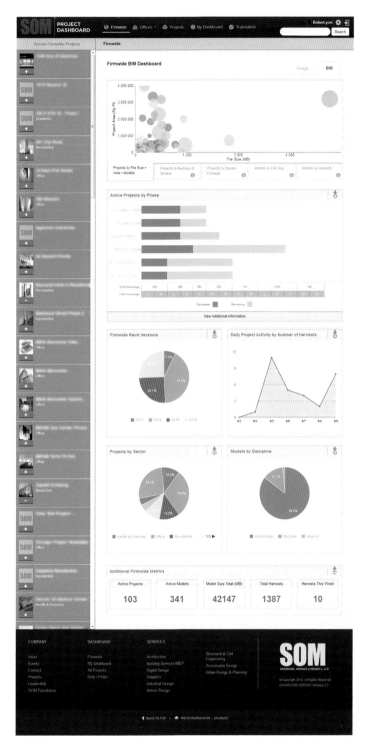

Figure 6–14 SOM BIM Dashboard front page.
Copyright © Skidmore, Owings & Merrill, LLP

Is it too early to consider the "lifecycle BIM" clients seem to benefit from as long as consultants and contractors cannot even figure out how to align the BIM they produce? What can BIM Managers do to overcome these issues that have become ever more apparent as BIM enjoys increased uptake across the construction sector?

AS THE CONSTRUCTION INDUSTRY STILL SEEMS TO STRUGGLE TO GET DESIGN CONSULTANTS TO HAND OVER QUALITY INFORMATION TO CONTRACTORS, DOES THE OPTIMIZATION OF BIM FOR FACILITIES MANAGEMENT REMAIN NO MORE THAN A PIPE DREAM?

Paul Nunn: It's not about delivering BIM to FM, it's about delivering FM-ready models. If we produce a good BEP with the geometry and data in a format that the contractor can use efficiently we will have more than is required for transfer or integration with an FM system. Our approach is not to mention FM but just specify what we need. Too many BIM Managers, including myself, in years past promoted FM as the "be all" of BIM and insisted it couldn't be achieved unless the client selected an FM system up front and specified all the data required. In fact we've found that's rubbish. A good model(s) used for commissioning has too much data and needs to be edited back for FM. Most projects already have a requirement for commissioning data collection in various formats which can be adapted if COBie is not specified.

Rob Jackson: There are many preaching the virtues of BIM to FM and there are some organizations with convincing case studies. There are also design and construction professionals blaming clients for the lack of demand for BIM to FM. However, in my opinion design and construction needs to get its own house in order first before charging into offering these additional services. For me the current focus should be on "perfecting" the design and construction process.

Ronan Collins: In the near term, this is a niche market opportunity which typically only applies to projects for owners with a large portfolio of buildings or properties and an advanced attitude as to how they operate the premises, maintain the systems within their facilities, and plan future development works. BIM Managers can implement processes, technologies, data formats, and the like but before they are called into action, the clients or owners, architects, contractors, and facility managers need to be aligned on how a project will be designed, constructed, and operated. The overall strategy may include energy-use targets and preventative maintenance approaches. For the foreseeable future, BIM Managers will be focused on the design and construction phases of a project.

Working According to Local Guidelines and Standards

"Focusing our research and development on open standards has allowed us a medium-sized UK architectural practice to build up a reputation for understanding these workflows. This has allowed us to speak at software events, that don't focus on our own authoring tool, as an advocate of open workflows that are relevant to all model authors and users of open data."

Rob Jackson (Bond Bryan Architects)

Figure 6–15 Examples of international BIM guidelines.
© **Dominik Holzer/AEC Connect based on BSI (UK), GSA (US), NATSPEC (AUS), and BCA (Singapore)**

One element associated with the expertise of any BIM Manager is his or her in-depth knowledge about local BIM policies. Whereas in early days of BIM an understanding about process and technology were their primary concerns, BIM Managers around the world increasingly need to engage with both local as well as national standards, guidelines, and other policies. In addition to these overarching frameworks issued by governments, government agencies, or councils, BIM Managers are now more and more also confronted with guidelines by industry bodies or other associations. Some standards have emerged based on work by user groups from within the industry; others are based on open formats for collaboration such as the IfC OpenBIM initiative promoted by buildingSmart. What all these efforts have in common is that they have a major effect on how any organization structures their BIM efforts. In some cases BIM guidelines simply provide suggestions on how to streamline BIM efforts, in other instances, they put forward concise (and often even legally binding) frameworks for operation.

In the case of the former, BIM Managers can draw from these standards in order to inform their strategy and align them with the operational processes within their organizations. In the case of the latter, BIM Managers need to be fully aware about all the implications of these binding standards on the way they structure work-flows and information handover on any project they are involved in. BIM Managers thereby carefully need to interpret the standards and develop a strategy that aligns requirements with a range of business and BIM Management processes from training, document and model setup, the definition of BIM Execution Plans, plus many more. In some cases, national guidelines and policies are set up to effect fundamental change within their local construction markets. As a consequence, they greatly influence the BIM strategy of an organization; and any updates/changes to those policies need to be monitored closely by BIM Managers who otherwise risk missing out on developments that may affect their organization's business. One such example is the UK Publicly Available Specification (PAS) 1192 and its various components. These specifications are likely to be woven into a British Standard and they are the foundation and key reference to the production of an international standard (ISO 19650—Organization of information about construction works and Information management using building information modeling).

No matter if any of the PAS 1192–related specifications apply within their local market, BIM Managers can benefit from their content when structuring information within their organizations and across projects. BIM Managers who excel in their work are not only familiar with these standards, but they truly understand their key concepts and use them in setting up and managing BIM on projects.

HOW DO THE GLOBAL BIM EXPERTS RECOMMEND WEAVING LOCAL AND INTERNATIONAL POLICIES AND GUIDELINES INTO DAILY BIM SUPPORT?

Rob Jackson: For us in the United Kingdom this is a straightforward question. Much of the United Kingdom is focused on the emerging "Level 2 BIM" standards, protocols, and processes. This includes PAS 1192-2:2013, PAS 1192-3:2014, BS 1192-4:2014, PAS 1192-5:2015, CIC BIM Protocol, Government Soft Landings, Classification and a Digital Plan of Work (dPOW). Of course this builds on the "Level 1 BIM" standards, such as BS 1192:2007.

Beyond 2016 we will begin to see the emergence of further standards, protocols, and processes to support "Level 3 BIM," branded as Digital Built Britain (DBB).

Our whole approach is to align with the standards fully. This is implemented in our documentation, templates, training, and checking processes. We also provide support to the consultants, clients, and contractors we work with on live projects in order to resolve issues. This collaboration allows us to resolve issues for our projects but also to learn and understand how to adjust our approach. Also any software issues are immediately fed back to the appropriate vendor for resolution.

As a practice we have also focused heavily on developing our approach around open standards including ISO 16739:2013 (IFC) and COBie. We are also working on integrating BCF (BIM Collaboration Format) into our processes. We believe that open international standards are imperative in the long term but we also use these on a day-to-day basis on live projects.

Paul Nunn: Oh to live and work in the United Kingdom! We are currently working in Western Australia, Queensland, Victoria, New South Wales, and the Northern Territory and each have a different set of State BIM requirements or guidelines and in the case of New South Wales different ones for each State Department. Our approach is generally to push the UK protocols wherever possible but we also borrow from the many other countries doing BIM. Again, it comes back to having sufficient industry experience to have the confidence to adapt a process and argue its benefits.

Ronan Collins: While the United Kingdom is currently pushing BIM in a very strategic way, a lot of countries and even agencies within states don't understand the implications of specifying the use of BIM or even how to get started. As a result there are numerous varieties of standards, specifications, guidelines, contracts, etc. out there. Luckily, they all adhere to some fundamental principles on how to apply BIM within an organization or on a project. After all BIM has been developed on a global scale by global technology companies and buildings are generally assembled from timber, concrete, steel, glass, and aluminum. So there are a lot of common fundamental good practice principles. The BIM Manager's starting point is to produce a BIM Project Execution Plan and it must clearly state the BIM objectives or uses, the deliverables to be produced, the staffing required, technologies to be used, and then cover specific issues such as Level of Development, Coordinates, Units, etc. The BIM PXP will obviously be localized but it could make use of the UK PAS 1192, GSA, Chinese, or other internationally established standards. In the absence of a project execution plan, client specification or even an applicable national standard, a BIM Manager will struggle to implement procedures and processes to control a design or construction team effectively. It doesn't matter which standard is followed, the critical issue is to put a robust plan in place.

Moving Forward (While Catching Up)

"The BIM Management market is getting quite crowded with just about every architect and engineer now claiming to be BIM Managers and in some cases providing BIM Management for free if given the design consultancy. So our focus is working further along the journey: working with the project FM managers and FM software companies to help them integrate BIM into the software and upskill ourselves as future FM managers."

Paul Nunn, General Manager BIM at PDC

Figure 6–16 University of Nottingham Advanced Manufacturing Building (AMB), Nottingham, UK.
© **Bond Bryan Architects LTD**

Any discussion about the BIM Manager's future role has to be put in perspective with the shortcomings of the current status of the industry. As much as there is clearly scope for moving forward, there is equal scope for catching up on previous BIM developments to make the adoption of BIM more balanced across the entire building lifecycle.

One development pointing toward the future is obvious: BIM will continue to become broader and more encompassing. Processes and standards in its support therefore need to consider an ever wider group of stakeholders and their particular interests. For BIM Managers this means two things: First, BIM Management needs to transgress predominantly profession-specific isolation and become more inclusive and open. Second (and stemming from point one), BIM Management will require those involved to focus on information flows and data management across a great number of stakeholders. With ever-more profound integration between information sources from all walks of life and the design, construction, and operation in the built environment, BIM will facilitate the point of interaction.

BIM Managers may respond to these challenges either by removing themselves from the bigger picture and focusing merely on in-house modeling processes, or by considering their role as Information Managers and Data Engineers of the future. Whereas the former is by far the easier solution, the latter will require them to expand their skills and learn about the overarching drivers that link technology to design within our society. Those who go down the easy path may maintain some relevance for the years to come. Those who risk to embrace change may future-proof their role and, at the same time, become part of what comes after BIM.

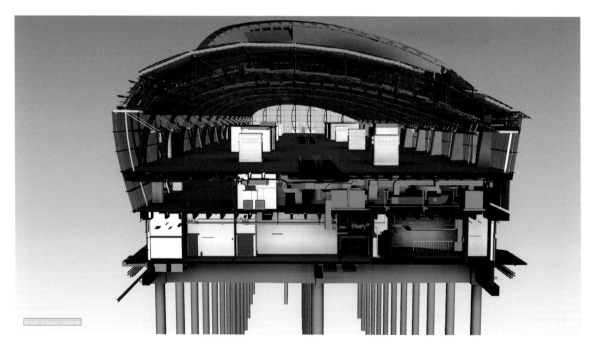

Figure 6–17 Airport terminal section.
© Intelibuild

BIM FUTURE FORWARD

Rob Jackson: I'm a bit of a cynic when it comes to the future. In the United Kingdom we have only had really four years to understand and adopt new workflows. We have less than 12 months until "Level 2 BIM" is officially required for UK government centrally procured projects. The technology is still not fully supportive of these new standards and processes (despite vendor claims) so this needs to catch up first. We must get the vast majority of the industry up to a certain level of BIM first before we can plough on.

Of course there are those who need to future gaze and develop new standards. I am fully supportive of this. However, there is still much work to get the whole industry to adopt BIM.

That said, for me the future is about open workflows. I believe passionately in this and without it I believe we simply have moved the silos from people to software. These open workflows need to be driven by customers. Sadly at the moment many only focus on their own tools and don't look beyond to how data will be used for other purposes.

Paul Nunn: I think it is about maintaining your knowledge and participating in industry forums, etc. We work closely with two different universities and one TAFE college participating in discussions around future BIM and helping them develop short-term and long-term BIM courses and integrating BIM into their existing courses. We also participate in as many industry BIM forums and committees as we can to understand what industry is looking for.

Ronan Collins: We all work in a conservative industry operating with long-term operational time frames and very traditional roles and forms of contract. Clients have budgets, architects and engineers compete for work by lowering fees, and contractors price jobs by taking account of risks. Unless and until we can find a collaborative form of planning, designing, building, and operating buildings, those procurement systems will prevail. Inherent in that environment are silos for different disciplines, a defensive culture to avoid blame, and a lot of abortive work. BIM can address some of the challenges but it's only a component part of the solution. A lot of other factors need to change.

Robert Yori: BIM goes way beyond skill levels in a particular tool or set of tools. At its core, it's about an informationally driven way of thinking about a project. In that light, we're already seeing what's next: computationally driven design, design-intensive analysis, and the closing of the design-fabrication gap. The industry is immensely broad in range, but if we compare it to the auto industry, we see there's room for that range. Custom coach builders, Ferrari, BMW, Ford, and Toyota all use technology, but do so differently from each other because they serve different markets and customer needs. It's used where it helps and not where it hinders. I see the same in a digitally mature AECO world.

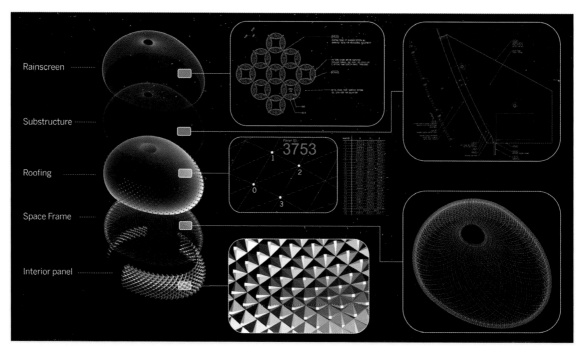

Rainscreen

Substructure

Roofing

Space Frame

Interior panel

Figure 6–18 Mosque shell fabrication.
© Skidmore, Owings & Merrill, LLP

Thank you to all the experts who so generously offered their thoughts and insights for this chapter: Wesley Benn of RTC Events Management; Ronan Collins of InteliBuild; Rob Jackson of Bond Bryan Architects; Paul Nunn of PDC; and Robert Yori of Skidmore, Owings & Merrill LLP.

Endnotes

1.   http://www.laiserin.com/
2.   www.aecbytes.com/blog
3.   http://extranetevolution.com/
4.   http://www.allthingsbim.com/
5.   http://www.lauraehandler.com/
6.   http://bimcrunch.com/
7.   http://practicalbim.blogspot.com.au
8.   http://collaborativeconstruction.blogspot.com.au/
9.   http://datadrivendesignblog.com/
10. http://bimblog.bondbryan.com/
11. http://bimfix.blogspot.com.au/
12. https://www.augi.com/
13. http://revitall.wordpress.com/

14. http://blog.graphisoftus.com/

15. http://www.archicadmonkey.com/

16. http://www.theprovingground.org/

17. http://www.theb1m.com/

18. http://www.meetup.com/NYC-RUG/

19. http://brisbim.com/

20. http://melbim.com.au/

21. https://twitter.com/sydbim

22. http://www.bimwest.org/

23. http://pages.info.bentley.com/events/

24. http://au.autodesk.com/

25. http://www.tekla.com/uk/conference-2015/index.html

26. http://www.bimmepaus.com.au/forum-current.html

27. http://rtcevents.com/index

EPILOGUE

These are exciting times for those who deal with the technology and process side of project design and delivery. More and more they are confronted with a shift in mindset where lifecycle aspects of building projects become a central part of planning and delivery. When applying BIM, the necessity of understanding and managing information flows across the supply chain from design, engineering, manufacture, and construction to the operation of built assets becomes apparent. Next to a realization about the interconnectedness of these previously often disjointed processes come the opportunities for scaling up BIM to the precinct, city, or regional level, producing information models that refer to an even broader geographic context. It is the intelligence of digitally interpretable data and the resulting value of data associated with BIM that will drive its continued adoption in the future.

BIM Managers are central to the delivery of projects and the advancement of the discourse related to BIM. They carry essential custodianship over innovation and process change facilitated by technology across a wide range of professions. Equally, they are the interpreters of guidelines, policies, and standards that influence on a broader level how BIM gets delivered on projects. On top of all of that, BIM Managers often act as the conduit of information within their organization and across the entire project team. They support and empower others who are less technically inclined, or those who simply prefer to focus on different sets of activities. In doing so, BIM Managers constantly strike a balance between the technically possible and the practically advisable. In some cases, this also requires BIM Managers to step back and acknowledge where their expertise ends and the knowledge of others is required. Construction as a process has become multifaceted and complex to the point where nobody can "know it all" on medium- or large-scale projects. What remains relevant though is to understand how information flows and what sort of information is relevant to the various stakeholders who collaborate on construction projects.

BIM is here to stay as a hot topic for at least a decade, depending on local differences and levels of uptake. During and beyond that period, it is likely that BIM will dissolve into a great number of activities associated with design, engineering, project management, and delivery. Aspects of BIM that affect ongoing Asset and Facility Management will establish themselves as project deliverables through feasibility studies and via project briefs by way of natural selection. If clients perceive a benefit of requesting particular services, they will do so and they will increasingly become better at articulating what exactly they expect from BIM.

For now, they still rely on the supply chain to help inform their deliverables on a project-by-project basis. Current developments point toward a future where authorities in an ever larger number of developed economies prescribe BIM as part of common project delivery methods on government-funded projects. BIM frameworks are already in place in some Scandinavian counties (such as Norway's Statsbygg), in the United Kingdom where authorities require BIM Level 2 on all public works, or in Singapore where the local Building and Construction

Authority (BCA) has implemented mandatory BIM e-submissions for regulatory approval. These are merely a few examples of an ever growing list.

Stepping away from the bigger strategic picture, there is still a lot to do and a lot to learn for BIM Managers. First, they need to become better in judging how their role and their output align with the core business of their organization. This is an essential step away from being a mere facilitator of technology use on the floor toward becoming a key member of an organization's leadership and advancement team. BIM Managers are typically confronted with accumulated needs when it comes to their skills in articulating concise business cases. Their ability to communicate clearly and concisely to their leadership what it is they require to make BIM work constitutes a second major step in their development. This is then followed by activities related to process-change and, the establishment of back-of-house protocols, the delegation of BIM-related workload, and the facilitation of a lively discourse about to technology uptake and dissemination. All of these parts play out differently depending on the experience level and the aptitude of the individual, as well as the immediate professional context in which they find themselves. The better they are at their job, the more likely they will manage to "work themselves out of it," which would be the ultimate achievement any BIM Manager should aspire to.

INDEX

Page numbers in *italics* indicate Figures; those in bold type refer to Tables.